THE GREAT CHAIN OF BEING

Borgo Press Books by BRIAN STABLEFORD

Algebraic Fantasies and Realistic Romances: More Masters of Science Fiction
The Best of Both Worlds and Other Ambiguous Tales
Beyond the Colors of Darkness and Other Exotica
Changelings and Other Metamorphic Tales
A Clash of Symbols: The Triumph of James Blish
The Cosmic Perspective and Other Black Comedies
The Cure for Love and Other Tales of the Biotech Revolution
The Devil's Party: A Brief History of Satanic Abuse
The Dragon Man: A Novel of the Future
Firefly: A Novel of the Far Future
The Gardens of Tantalus and Other Delusions
Glorious Perversity: The Decline and Fall of Literary Decadence
Gothic Grotesques: Essays on Fantastic Literature
The Great Chain of Being and Other Tales of the Biotech Revolution
The Haunted Bookshop and Other Apparitions
Heterocosms: Science Fiction in Context and Practice
In the Flesh and Other Tales of the Biotech Revolution
The Innsmouth Heritage and Other Sequels
Jaunting on the Scoriac Tempests and Other Essays on Fantastic Literature
The Moment of Truth: A Novel of the Future
News of the Black Feast and Other Random Reviews
An Oasis of Horror: Decadent Tales and Contes Cruels
Opening Minds: Essays on Fantastic Literature
Outside the Human Aquarium: Masters of Science Fiction, Second Edition
Prelude to Eternity: A Romance of the First Time Machine
The Return of the Djinn and Other Black Melodramas
Slaves of the Death Spiders and Other Essays on Fantastic Literature
The Sociology of Science Fiction
Space, Time, and Infinity: Essays on Fantastic Literature
The Tree of Life and Other Tales of the Biotech Revolution
Yesterday's Bestsellers: A Voyage Through Literary History

THE GREAT CHAIN OF BEING

AND OTHER TALES
OF THE BIOTECH REVOLUTION

by

Brian Stableford

THE BORGO PRESS

An Imprint of Wildside Press LLC

MMIX

Copyright © 1993, 1994, 1995, 1996, 2008, 2009 by Brian Stableford

All rights reserved.
No part of this book may be reproduced in any form without the expressed written consent of the publisher.

www.wildsidepress.com

FIRST EDITION

CONTENTS

About the Author 6
Introduction 7

Following the Pharmers 11
The Unkindness of Ravens 48
The Great Chain of Being 53
Sleepwalker 72
The Beauty Contest 76
Burned Out 100
Inherit the Earth 126

ABOUT THE AUTHOR

BRIAN STABLEFORD was born in Yorkshire in 1948. He taught at the University of Reading for several years, but is now a full-time writer. He has written many science fiction and fantasy novels, including *The Empire of Fear*, *The Werewolves of London*, *Year Zero*, *The Curse of the Coral Bride*, and *The Stones of Camelot*. Collections of his short stories include *Sexual Chemistry: Sardonic Tales of the Genetic Revolution*, *Designer Genes: Tales of the Biotech Revolution*, and *Sheena and Other Gothic Tales*. He has written numerous nonfiction books, including *Scientific Romance in Britain, 1890-1950*, *Glorious Perversity: The Decline and Fall of Literary Decadence*, and *Science Fact and Science Fiction: An Encyclopedia*. He has contributed hundreds of biographical and critical entries to reference books, including both editions of *The Encyclopedia of Science Fiction* and several editions of the library guide, *Anatomy of Wonder*. He has also translated numerous novels from the French language, including several by the feuilletonist Paul Féval and various classics of French scientific romance.

INTRODUCTION

This is the sixth collection of "stories of the biotech revolution" that I have published, the others being *Sexual Chemistry* (Simon & Schuster UK, 1991), *Designer Genes* (Five Star, 2004), *The Cure for Love* (Borgo Press, 2007), *The Tree of Life* (Borgo Press, 2007) and *In the Flesh* (Borgo Press, 2009). There have also been seven novels of the same ilk: *Inherit the Earth* (Tor 1998), *Architects of Emortality* (Tor 1999), *The Fountains of Youth* (Tor 2000), *The Cassandra Complex* (Tor, 2001), *Dark Ararat* (Tor, 2002), *The Omega Expedition* (Tor, 2002) and *The Dragon Man* (Borgo Press, 2009). The majority, but by no means all, of the stories share a common future-historical background, an early version of which was first sketched out in a futurology book, *The Third Millennium: A History of the World 2000-3000*, written in collaboration with David Langford and published by Knopf and Sidgwick & Jackson in 1985. The series will continue for as long as I can see and type, and for as long as there are any publication outlets left (although I have to admit that all of those possibilities now seem to be direly short term prospects).

The foreseeable future tends to date even more rapidly than frail human flesh, and any futurological project that does not see some of its more cherished hopes, paranoid fears and confident anticipations betrayed by twenty-five years of actual history cannot have risked much in its initial assertions, but the broad shape of the future trajectory mapped out for global society in this series has proved sufficiently robust to stand the test of time without shattering completely. The idea that the world is inevitably heading for a worldwide self-inflicted disaster, compounded from the ecocatastrophic effects of population growth and the climatic spin-off of the misnamed "greenhouse effect" seemed like a safe bet in 1984 and is now beyond the reach of any residual skepticism. The one thing that early versions of the scenario got wrong was the speed with which the disaster would unfold; it now seems so urgent and imminent that the second, and more speculative, part of the initial prospectus—that once the "Crash" has done its worst, the combination of bitter experience and new technologies might permit the construction of a better and more Utopian global society—now seems desperate in its optimism.

Although the original version of the scenario extended over a thousand years, and had no alternative to painting in broad strokes, I took the decision when I began the series extrapolated from it to place most of the stories in small-scale domestic settings, and to depict future biotechnologies, wherever possible, as aspects of everyday life, working massive changes subtly and by stealth rather than by means of spectacular narrative twists and explosions. This was, of course, a fatal decision in terms of the series' potential popularity within a genre that thrives on cheap but flashy melodrama, and helps to explain why such stubbornly low-key stories as "The Beauty Contest" never sold, while the cynically over-melodramatized "Inherit the Earth" proved so appealing that when a publisher finally consented to do the series of novels (ten years after the proposal was initially put on the market) he insisted on issuing the books out of chronological order so as to start with the expanded version of that one.

Two of the stories here are very short, and the ideas might have been better displayed if they had been deployed over the 6,000 words that is nowadays compulsory for the "theme anthologies" that still continue to cling on to a tiny corner of the devastated marketplace, but the simple fact is that in days gone by there was often more demand for short stories than long ones, for various practical reasons. In the 1990s, when I used to submit work regularly to the British magazine *Interzone* my work was routinely rejected on the grounds of being too long, and if ever I happened to mention work that I was currently doing to David Pringle, the editor, he was prone to say: "For God's sake don't send it to us—we simply can't handle material of that length." This never prevented him, however, from occasionally ringing me up and saying: "We've got a very long story in the next issue and we need some very short ones so that we can fill up the contents page—could you knock off a couple for us?"

The present collection is named for "The Great Chain of Being" partly because that is my favorite among the stories it contains, partly because it is the most imposing title, and partly because it is by far the most optimistic story in the set. Two thousand years of Christian fantasy have not managed to produce a single image of a Heaven to which anyone would actually want to go, if they had the choice, and some attempts have been so perversely horrific as to make the less intimidating fraction of the imagery of Hell seem quite attractive, so my chances of doing any better seemed slim at first, but in the end, the task did not seem unduly difficult. On the other hand, such challenges inevitably come down to a matter of taste, and it is conceivable that mine is a trifle eccentric.

"Following the Pharmers" first appeared in the March 2008 issue of *Isaac Asimov's Science Fiction*. "The Unkindness of Ravens" first appeared in *Interzone* 90 (December 1994). "The Great Chain of Being" first appeared in *Future Americas*, edited by John Helfers, published by DAW in 2008. "Sleepwalker" first appeared in *Interzone* 105 (March 1996). "The Beauty Contest" appears here for the first time. "Burned Out" first appeared in *Interzone* 70 (April 1993). "Inherit the Earth" first appeared in the July 1995 issue of *Analog*. As indicated in the list above, an expanded version of the last-named novella was published in book form by Tor, but the ending of that version was extensively rewritten to editorial instruction, so the two works are markedly distinct.

FOLLOWING THE PHARMERS

"When tillage begins, other arts follow. The farmers, therefore, are the founders of human civilization."
—Daniel Webster

It was early in June that the antheric alates began appearing on my verandah. At first I assumed that they were natural insects—some new species of miniature butterfly nurtured in the evolutionary hothouse that the Holderness had recently become. Their tiny wings were brightly-colored, with a quasi-metallic sheen that enabled them to flare like sparks in the bright light of noon and twinkle like stars in the evening, when the sun sank into the bosom of the Wolds. Initially, I welcomed their arrival as a fortunate discovery, a safe distraction from the burdensome aspects of my isolation.

Once I had examined a couple of the motiles with a magnifying glass I realized that they weren't insects, but I was still possessed by the idea that they might be some new kind of invertebrate animal—perhaps an entirely new branch of the arthropoda, spun off by bold mutation from one of the many former sea-creatures that were adapting with astonishing rapidity to the Yorkshire Everglades. Once I had put one under the microscope, though, I realized that they were vegetal, and also that they were artificial.

That was when I started cursing. It meant that I had a new neighbor. The whole point of our moving to Hollyn—a place that wasn't even supposed to exist any more, in the official cartography of New England—had been to give us the opportunity to do our work in peace. I hadn't wanted neighbors when Marie was still around; I certainly didn't want one now that she was gone, unable to return.

I wasn't completely isolated from human contact, of course, but I didn't count the Patrington communards as "neighbors". They performed a useful intermediary function in transmitting my produce to the wholesalers in Hull—a necessary function, given the amount of chemical assistance I'd have needed to go all the way to the city on my own behalf. In any case, Patrington, which had also benefited from an unexpected and so-far-unrecorded re-emergence from the shallows of the Holderness to become a substantial new island, was a good seven kilometers away. The alates, I judged, must have come from somewhere considerably closer.

The communards were small pharmers like me; they planted, nurtured and processed their crops according to strict chemical rituals, never taking the risk of producing anything new. Whoever was producing plants with alate pollen-sacs, on the other hand, had to be an artist: an innovator of considerable daring as well as abundant talent. From the viewpoint of a small pharmer, artists qualify as loose cannons: mad, bad and dangerous to have around. I knew, because I'd fancied myself as a bit of an artist in the days of my *folie à deux* with Marie, and even before that, in the days when we had both been wage-slaves in one of the corporate giants making up Big Pharma.

My verandah faced north, in order to give me shade from the hostile UV of the noonday sun, and that was the direction from which the alates were coming. There shouldn't have been anywhere in that direction for them to come from, but I knew that if the stubborn ancient walls of Hollyn and Patrington could provide the foundations for marvelous growths of littoral limestone, and hence for newborn islands where plastishacks could be securely bedded, there was every possibility that parts of Withernsea could do likewise.

Of all the former dwelling-places in the Holderness, Withernsea was the one that generated the most legendary echoes—far more than Hornsea, which had been a considerably bigger town in the Ice Age. As its name proudly declared, Withernsea had been on the coast in those days, and would now be unsheltered on its eastern shore from the full wrath of the North Sea storms—but what it lacked in safety it might make up in romance, at least in the eyes of an artist.

Withernsea was a lot closer to Hollyn than Patrington, as any sort of creature might fly, but I had no idea whether there was a navigable channel through the algal dendrites that reared up from the new sea bed, whose colonization of the Holderness grew more insistent with every year that passed. I never went out in the motor-boat for "leisure purposes", but if I ever had I would have headed vaguely eastwards, in the direction that would have qualified as "inland" before the old Ice Age land had been gradually swallowed up by the salt-marsh.

I considered the possibility of ignoring the matter, simply hoping that it wouldn't become a problem. If I had been a fungal specialist, like the communards of Patrington, that would have been a justifiable strategy, but I wasn't. I had three species of flowering plants producing reliable cash crops. The rape and the poppies were safe enough for the time being, but there was no way of knowing how far across the angiospermal spectrum the artist's experiments might eventually range, and the foxgloves might already be in hazard. Pharmed foxgloves are notoriously vulnerable to what the technical jargon terms "bizarre pollination", in spite of the insect-repellents built into their nectar. I didn't suppose for a moment that those inbuilt insect repellents would have the slightest effect on antheric alates.

For that reason, I really needed to talk to my new neighbor about the situation, if I could. With luck, all I'd have to do would be to ask him politely to tighten up his security-measures, and he'd be willing to oblige. There is, after all, a certain code of politeness involved in living outside the law; no one with any sense wants to give anyone else too powerful a reason to stir up trouble. I'd have to get my head in condition to make the trip, but I trusted my own products and visiting wasn't something I'd ever had to do with sufficient frequency to risk another hook.

I didn't know how long it would take me to find a viable route to Withernsea, but I didn't dare set off in the early morning, even with a canopy over the boat and the shade of the algal dendrites to limit my UV exposure. Given that it was June, when the days lasted far longer than the nights, the prudent thing to do was to pop the requisite pills in late afternoon and set off in the right direction, establishing a deadline for the search that would guarantee me a safe passage home before the twilight dwindled away.

I didn't make it on the first day, but I figured out a mazy route that got me close enough to the re-risen Withernsea not merely to estimate the contours of the island but actually to glimpse the roof of the largest of the plastishacks in which the artist had set up production. It was hard to miss, not only because of its capacious size and flamboyant architectural design, but because of its blatant disregard for the most elementary camouflage. If a copter were ever to fly over my place, its pilot would need a keen and attentive eye to make it out, but the new building stuck out from its surroundings like a tarantula on a lacy net curtain. The Hull police had far more urgent things to do at present than explore the Holderness, which lay outside their jurisdiction, but the boldness of the new development was still reckless.

The next day, I followed the mazy path I'd already mapped out with all possible speed, starting at four-thirty, and had found a way to the shore of the new island by six. I tied the boat up in the shade of a mock-willow, and made my way stealthily over the virgin coraloids to the sturdy platform on which the complex of plastishacks had been erected. The central element, at least, was more mansion than shack.

Ever since the Great Migration had begun, technologies for erecting instant houses had been subject to tremendous selective pressure, forcing them to evolve with the same tachytelic fervor as the new littoral ecosystems that were recolonizing the drowned land. Even so, the house seemed to institute a significant step forward. I'd never seen anything like it advertised on TV. The artist was obviously an exceedingly rich amateur rather than the kind of impoverished optimist who was accustomed to starving in the proverbial garrets of the Ice Age.

My heart sank as I looked at the place from a distance, sheltered by the algal undergrowth, and I nearly lost my nerve. In spite of the chemical fortification, I wanted to change my mind, turn around and go back home. I knew, though, that if I did that I'd eventually have to come back, probably sooner rather than later. The combination of necessity and curiosity was just powerful enough to give me the courage to continue going forward and knock on the door. My approach was tentative, as much for fear of guard dogs or an entourage of bodyguards as my native inclinations, but the place was utterly quiet. If there was anyone at home, they were busy about their daily toil.

I knocked, and waited.

* * * * * * *

The person who answered the door was a casually-dressed female, whose apparent age was about twenty-one. I didn't immediately jump to the conclusion that she was a lowly servant, though. The kind of wealth necessary to buy a mansion-sized plastishack could also buy a great deal of cosmetic somatic engineering, and I assumed that even rich people dressed casually when they weren't expecting visitors. The woman might easily be the kind of apparent twenty-one-year-old who'd been around for more than a century. Those kinds of people, so rumor had it, often went in for exotic hobbies.

"I'm Daniel Anderson," I told her, while she looked me coolly up and down. "I'm probably your nearest neighbor, unless there's someone closer to the north."

"I'm not supposed to have any neighbors," the woman replied, her use of the personal pronoun suggesting that she was the artist herself rather than any mere hireling. "That was the whole point of moving out here. There isn't supposed to be anyone living between here and Hull. There isn't even supposed to be anywhere for them to live." The way she stood in the doorway was manifestly imperious; she was definitely the mistress of the house.

"This place isn't supposed to exist either," I pointed out. "My smallholding is on new ground that formed above the walls of the church at an Ice Age village called Hollyn—the steeple came down and the roof caved in, but the rest stood firm. I run a small pharm there."

"How nice," she said. She still hadn't told me her name, let alone invited me in. "What do you grow there?"

"Psychotropics. Mostly amanita and muscaria derivatives, some opiates, a few exotic oils and digitalids."

"Digitalids?" she queried. "Does that include inspirationals and focal intensifiers?" She was definitely an artist.

"Yes it does," I said. "Nothing very exotic, though—standard stuff you could buy off the shelf if the law were a little saner and Big Pharma a little less paranoid."

"Ah," she said. "You've found some of my alates, haven't you? You're worried about the possibility of transgressive cross-pollination. How far away in your pharm, exactly?"

"A little less than three kilometers, as the alate flies," I told her.

"That far? I had no idea that my little treasures had that sort of range, even with the aid of a favorable wind. The wind mostly blows from the west, carrying escapees out to sea, but the land and sea breezes are brisker in summer, and they alternate with a certain forceful regularity. I can assure you that my alates pose no danger to your poppies and foxgloves. For the moment, I'm only working with roses, lilies and orchids, and I've no intention of broadening my experimental range in the present phase of my campaign."

"How long is the present phase of your campaign likely to last?" I asked, lending a slight ironic emphasis to the odd phraseology.

She didn't answer. She didn't shut the door in my face, though. She was obviously intrigued to discover that she had a neighbor within a mere three kilometers. She wanted to know more about me. She knew full well that it would be easier and safer to do that at a distance, but she was apparently the kind of person who preferred operating directly, face-to-face—unlike me.

"What do you think of my designs, Mr. Anderson?" she asked.

I didn't suppose that she cared about my critical opinion. She was fishing for information about my politics, and the extent of my biochemical expertise. "It's not a matter of aesthetic admiration, so far as I'm concerned," I told her. "I'm sure that the natural flowering plants that are busy colonizing the New Everglades are too discriminating to entertain foreign pollen, but the whole point of engineered flowers is to welcome hybridization and facilitate eclectic recombinations. It's hard enough keeping my poppies and foxgloves from unnatural intercourse with one another, without having dozens of varieties of ambitious pollen flying in of their own volition. Would it be possible for you to tighten up your containment procedures? Not so much for my benefit as for your own—it's only a matter of time before other people begin finding your stray produce."

"I can assure you," she said, "that the police aren't going to bother me here." She sounded very confident. She looked me up and down again, as if measuring me for aggressive potential. I had to admit that, from her point of view, I might easily seem dangerous, no matter what kind of subtle defenses her fancy house was fitted with or how many other people would come running in response to a cry of alarm. After a suitable pause, though, she nodded and moved aside, inviting me to come in.

I hesitated. I wanted to turn and run, assuring myself as I went that I'd done what I came to do, and that there was no need to string it out.

She frowned, obviously having divined the impulse, and finding it rather unflattering. "I'm Judith Hillinger," she said, as if that were guaranteed to settle the matter.

It took a couple of seconds for the reflex to kick in and bring the memory to the surface. The moment of realization must have been clearly legible on my face. "Please come in," she said, to complete her victory. "Given that we're neighbors, we ought to get to know one another a little better."

She showed me into a room that the mansion's architect must have envisaged as a "reception room", even though the edifice was located in a place to which invited guests and stray callers would have to make a long and awkward journey. I sat down on a settee, which was upholstered in fancy leather that had never been worn by a cow, and accepted her offer of a glass of iced water.

"You have the advantage of me now," she said. "You probably know my entire life-story, up to the point when I was released from jail. Even if you somehow contrived to miss the scandal, you can extract every detail from web archives in a matter of minutes. I know nothing at all about you, though, and if I were to feed your name into a search engine I'd probably find it very difficult to sort out one particular Daniel Anderson from all the rest."

I knew how slight and short-lived any advantage I might possess would prove. If she wanted to find out everything there was to know about me, she could do it—except, of course, for the one thing that nobody knew. I didn't even have the momentary advantage of still being familiar with the lurid details of her case, All I remembered for sure was that she was sufficiently well-connected to have got away with a slap on the wrist for the kinds of flagrantly illegal but essentially unhazardous plant engineering she'd been doing a decade or so ago, and that she had inherited so much money from her late father that the maximum fine would hardly have made a dent in her fortune. Instead, she'd elected to go to trial, and had turned the courtroom into a media circus, making impassioned speeches in defense of the freedom of creativity, and the urgent necessity of humankind becoming the true masters of evolution.

I'd thought that what she'd done was foolish and counterproductive even at the time, when I too had still been enthusiastic to invent, innovate and become a master of mental evolution. She'd posed as a hero, but she was really just a nuisance, making it harder rather than easier for those of us who were content to work patiently in the shadows.

"There's nothing about me to interest someone like you," I told her, not knowing whether to hope that it was true. "I'm just a pharmer, trying to make a dishonest living in peace."

"Which implies, I presume, that you haven't got a criminal record—yet."

"No. Are you going to turn me in? I suppose that a simple phone call from you would be enough to bring police copters scurrying from Hull to Hollyn, no matter how much they have on their plate."

"Don't be ridiculous," she said. "I'm in hiding, just as you are—and for me, as you'll understand, that's a little more difficult. The police won't bother me, as I said, but that doesn't mean that you couldn't cause trouble for me. If you were to tip off a certain section of the media...."

"I wouldn't," I said. "As you say, it's as much in my interest as yours to be discreet—which is why I'm here, to warn you about the alate problem. I just want us to be good neighbors. As far as I'm concerned, you have as much right to be here as I have, and to do exactly as you please—but I need to protect my investment. My margins are a trifle thin right now; the market's oversupplied, and the dealers I work with have troubles of their own."

"Perhaps you ought to be developing new products," she said.

"I've tried that," I admitted, trying to keep the bitterness out of my voice. "Research in psychotropics is difficult and dangerous; testing new products is the sort of thing that can seriously damage your thought-processes. Adventures of your sort carry far less hazard, and the results are much easier to evaluate. I presume that you don't have to worry overmuch about regulating your turnover and protecting your profit margins."

"You're right, I suppose," she said, insouciantly. "Given your apparently-straitened circumstances, I dare say that it would be difficult for you to relocate—and why should you, given that you were here first? You're right; the best thing is for us to make the effort to be good neighbors. If my alates' range extends to several kilometers, I'll have to make more effort to contain them. It'll be a nuisance, but it's not impractical. I originally intended to surround the compound with high fences—nothing obtrusive, just spidersilk mesh sustained by discreet poles—but I let it slide when I discovered how difficult it is to make them storm-proof. I'll just have to steel myself to the necessity of making frequent repairs. Maybe I can get away with shielding the southern and eastern sides, so that any fliers that go a-wandering will be lost at sea. Will you give me a little grace, so that I can experiment with potential solutions until we find one that suits us both?"

"That's fine," I assured her—and it was, indeed, a better deal than I had any right to hope for, given that she probably had enough money to force me out or crush me like a bug, if that had been the way her instinct worked. "Thanks—I appreciate it. If there's anything I can offer by way of trade..."

"Of course there is," she said. "It will be handy, now I come to think about it, to have a local supplier. I'll pay you the retail price, and I'll trust you not to poison me."

The last comment was probably more of a threat than an expression of confidence, but she said it so lightly that I didn't take offence.

"Are you living here alone?" I asked.

"Oh no," she said. "I don't pay anyone to answer the door, because I didn't expect to have any uninvited visitors, but I'm not alone. I have three technical assistants, a cook-housekeeper and a boatman. The cook and the boatman are out fishing at present. Do you live alone, Mr. Anderson?"

"Yes," I said, shortly.

"That must be rather lonely," she said. "Perhaps you might come to dinner some time—but I'd really rather that you didn't drop in uninvited, if you don't mind. I'll give you a number to call, and you must give me yours. Would you like to see my laboratory?"

The last sentence came as a complete surprise, given that its immediate predecessors had implied that I was being brushed off now that our business was settled. She was an artist, though, and I was a pharmer who'd confessed to having done original work in the past. She had some reason to expect that I'd be capable of understanding her lab-work and appreciating its results. I also figured that she probably wanted to satisfy my curiosity, so I wouldn't have quite so much incentive to come back again.

"Yes," I said. "I would."

* * * * * * *

The lab was impressive, as it had every right to be, given the money that had obviously been lavished on it. Judith Hillinger's three technical assistants were equally impressive, at least to look at. Not one of them looked a day over twenty-one, although I guessed that they'd all had help in that regard. All three were female, though, so their expensive looks were presumably going to waste, unless the absent cook-housekeeper and boatman were both male and similarly cosmetically enhanced. I felt very old and very ugly, and I wasn't at all reassured by the politely disdainful way the three women looked at me as they were introduced, one by one.

I made suitably complimentary murmurs in confrontation with the genomic analysis kits, the chromosomal maps and the batteries of restriction enzymes, although the only thing that really impressed me was the sophistication of the proteonomic analyses. I made a similar show of being impressed by the seed nurseries and the hydroponics. I didn't have to make any effort at all, though, when we finally went into the greenhouses, whose contents simply took my breath away.

As Judith Hillinger had told me, she was working with roses, lilies and orchids—all long-time favorites of floral engineers. The blooms themselves didn't look particularly beautiful and unusual, by the standards that had been established half a century ago, and the symphony of their nectar was also expectable—but it wasn't the shape, color and scent of the blooms that stunned me with amazement.

I'd already seen the alates, of course, perched in twos and threes on the rail of my verandah, or fluttering in mid-air in tiny flocks of six or seven—but those were escapees a long way from home. In the greenhouses, the air was filled with them, not merely in their thousands but in their hundreds of thousands. It was their riotous colors, and the play of light on their wings, that struck me with extreme aesthetic force.

As Judith Hillinger moved among them, the alates settled on her body in their hundreds, and she adjusted her movements so as not to risk crushing them. I did likewise, and as we passed through the greenhouses we both seemed to be moving in slow motion, having undergone a metamorphosis into something far richer and stranger than anything merely human.

She had to put a hand over her mouth to shield it from invasion in order to speak, but she had a speech to make and she wasn't about to be inhibited.

"This is the way it should have been, Mr. Anderson," she said. "This is the path that evolution should have taken. This is one of the reasons why we must become masters of evolution as swiftly as possible—to correct the errors of natural selection. We'll have to start with the harmless ones, of course, in order to establish the principle—but pretty little ventures of this kind will only be the beginning."

This prompt allowed me to remember a little bit more about the content of the ostentatious speeches that Judith Hillinger had made in court when she'd tried to make herself a martyr for the creationist cause. She'd compared the work of natural selection to that of early computer programmers, who had been far more interested in finding a way to get the job done than in writing elegant code. As computing power and computer networks had grown at an explosive rate, all kinds of hasty improvisations had been built into source-codes, their initial weaknesses compensated by an ever-increasing mess of ungainly patches—which kept the whole thing working, after a fashion, but whose sheer mass and complexity prevented anyone from ever going back to basics and redesigning the code more efficiently and elegantly. By the same token, she'd argued, the ecosphere had blithely preserved anything that worked, however inelegantly, and had built up whole ecosystems by adding patches as they were thrown up by mutation—resulting in a vast ungainly complex that no one with any aesthetic intelligence would ever have designed, but which couldn't be comprehensively overhauled.

"When flowering plants first evolved," I said, to demonstrate to her that I was no fool, in spite of my exceedingly plain appearance, "the gymnosperms they were replacing set a very low standard of competition, in terms of their methods of pollination. The new forms didn't need to be very clever—just clever enough. It happened to be the evolutionary era in which the insects were undergoing *their* first major adaptive radiation, and insect pollination was good enough to do the trick. It would have been so much more elegant—albeit considerably more energy-expensive—for the angiosperms to invent pollen that could fly rather than rely on insects to serve as vectors, but the quick fix took hold. Once it had taken hold, the angiosperms and the insects became the major selective forces shaping one another's consequent evolution, so the whole ecosystem grew more and more elaborate, accumulating all manner of improvisatory patches—and the mutual success story was so spectacular that the prospect of going back to square one and finding a more elegant solution to the pollination problem vanished into the mists of possibility. Until now. You're not just trying to make prettier flowers for the home and garden, are you, Ms. Hillinger? You're trying to lay the groundwork for a whole new phase of plant evolution. So why start with roses, lilies and orchids?"

"I may be rich, Mr. Anderson," she said, "but I'm not *super*-rich. I need marketable products and healthy profits to finance further investment. This is just the beginning, as I said—and I'm not just talking about building a commercial empire."

"You're even more determined to get the law changed now than you were before you went to jail," I said, glad to be able to demonstrate that I was keeping up with her. "This is phase two of the great crusade, whose furtherance will be *seriously* expensive. It's not just a matter of buying more kit, hiring more techs and passing a few more brown envelopes to the Hull Police. Changing the law requires a war for hearts and minds, involving powerful advertising campaigns and relentless lobbying. Well, I wish you luck, Ms. Hillinger, I really do."

"Thank you, Mr. Anderson. I might be able to use a man like you, you know—and I could certainly use your pharm as a second experimental base. I think we could put together a very attractive package for you, which would put an end to your financial difficulties for some time to come, if you didn't want to stay on in a managerial capacity."

"I'm sorry, Ms. Hillinger, but that's out of the question," I said. "If I wanted to work for someone else, I'd never have quit Big Pharma."

"I'm not Big Pharma," Judith Hillinger stated, as if I'd just delivered a mortal insult. "I'm the absolute opposite. I'm starting out small, but I intend to become one of the leaders of the Revolution."

"If I weren't a confirmed loner, I wouldn't be holed up in the remoter regions of the Holderness," I told her. "I really do wish you the best of luck—but I'm just a pharmer, not a revolutionary. I don't want to be a part of your grand plan."

"You could get your looks fixed," she said, as if that were her idea of an offer that no one could refuse.

"I'm sure I could," I said, "but I think I'd rather wait for ugly to come back into fashion. I'm grateful, but the answer's still no. Can we just be good neighbors?"

She flashed me a smile that might have been intended to remind me exactly what I was turning down. "Of course we can," she said. "I'm sure that we shall."

* * * * * * *

When I got home, I found that I'd had visitors. I say "visitors" because they didn't seem to have been burglars, exactly, and they didn't seem to have been vandals, exactly. They'd messed things more than up a little, and they'd stolen some trivia, but they hadn't smashed anything up so badly that it would be difficult to make repairs, and they hadn't taken anything that I couldn't do without. Whatever their primary motive had been, it hadn't been robbery or destruction.

It occurred to me almost immediately, of course, that there might have been another reason why Judith Hillinger had invited me to look over her laboratory and her specimen-houses rather than letting me go home once we'd made an agreement. She had kept me there for a good two and a half hours after I'd told her where I lived and exactly how far away it was. I hadn't seen her make any phone calls, but I hadn't had my eyes on her all the time while I was being introduced to her three lovely assistants and shown around the labs. If the cook-housekeeper and the boatman had been fishing in the marsh rather than the open sea, there had been plenty of time for them to locate my pharm, take a good look around, and leave me abundant evidence that they'd been there.

I knew that I had to be careful about jumping to conclusions of that sort, because the pills I'd taken to enable me to make the excursion were notorious for inducing paranoid side-effects, but a pharmer has better reason than most people to bear in mind the old adage that just because you're paranoid, it doesn't mean they aren't out to get you. If it hadn't been Judith Hillinger's people, who could it have been? If the troubles the dealers in Hull were currently experiencing had extended backwards along the supply-chain, my visitors would surely have done a great deal more damage.

Whoever they were, the invaders hadn't done anything serious, but they'd left me a clear enough message as to what they *might* have done, had they been so minded. I knew that if I picked up the phone and told Judith Hillinger what had happened she'd be full of sympathy, and would put on a big show of being deeply hurt if I suggested, however delicately, that she might have had something to do with it. There was no point in doing that. After all, even if she had been responsible, she wasn't leaning on me hard—not yet. She wasn't trying to get rid of me, or force me to sign on to her Great Crusade. It was probably just that her idea of being a good neighbor wasn't quite the same as mine. She was probably prepared to play nicely, provided that it was perfectly clear who had the upper hand in the game and the power to crush the opposition, should the need arise.

I tidied up, and got on with my work.

For the next few days the numbers of the stray alates declined steadily; after a week had elapsed it became rare to see even one in the course of a day. Judith Hillinger had obviously instructed her hired help to put up some efficient netting to the south of her house. I was duly grateful for that, and tried to put her out of my mind. I didn't call her, and I didn't expect her to call me. On the afternoon of the first of July, though, my pocketphone trilled and when I interrogated the display I recognized the number she had given me.

"Ms. Hillinger," I said. "How nice to hear from you again. What do you need?"

"That depends on what you have for sale," she said.

"You mentioned inspirationals and focal intensifiers when I visited you," I reminded her. "I have basic products in both lines, as well as the usual range of memory-enhancers, euphorics, narcotics, hallucinogens and stimulants. I don't deliver, though—you'll have to send your boatman to collect the package."

"I'd prefer to collect them myself," she said, lightly. "It would get me out of the house for a while, and I'd be interested to look over your pharm. I showed you mine, remember."

"I remember," I assured her. "You'd be very welcome. I still need to know what you need, though, so I can make up a package."

"I'll make my selection when I've looked around," she told me. "I don't mind waiting while you assemble the package. I'll be there in an hour or so, if that's convenient."

"Do you need directions?" I asked innocently.

"I'm sure that my boatman can find you," she countered. "We have an Ice Age map."

An hour later, at five o'clock or so, her boat arrived at the crude jetty where I kept my own motor-boat tied up. Unsurprisingly, her boat was a lot bigger than mine, with a much nicer canopy to keep the UV at bay—which hadn't prevented Judith Hillinger from carrying a pink parasol, or deterred the boatman from wearing shades with skintight lenses the size of brandy-schooners.

The boatman was an exceptionally handsome man with an unfashionably muscular body; he could have broken me in half with his bare hands while smiling like a model for the latest generation of smart underclothes. Judith Hillinger introduced him as Jacquard, but he stayed with the boat while I led her to my home. The shack had never seemed more deserving of its name.

I gave her the tour, all too well aware of the fact that the plastic-shelled igloos sheltering my poppies and foxgloves were the merest shadow of her magnificent greenhouses, and that my mushroom-cellars were hideously rank by comparison with the nectar-laden air of her entire establishment. She was very polite, except when I showed her what had once been my research lab, where Marie and I had tweaked psychoactive compounds in search of something far more radical than the palliative treatments for Asperger's syndrome that Big Pharma had commissioned us to develop.

"You've let this part of your work languish, I see," she observed. "You shouldn't have done that, Mr. Anderson. A proteonomicist like you ought to be working at the cutting edge, not growing standard products—isn't that why you went out on your own in the first place?"

She'd obviously taken the trouble to sort me out from all the other Daniel Andersons, and had been at least a little bit intrigued by what she'd found. She must have known that I hadn't gone out of

Big Pharma "on my own", but she was carefully refraining from mentioning Marie, at present.

"I went into psychotropic proteonomics because it was fashionable," I said, modestly. "I was no hot shot. Everybody working in psychotropics at the time dreamed about discovering the ultimate high, so I bought my ticket in the lottery. It turned out to be a loser. I'm just a small pharmer now—it's not glamorous, but my products are guaranteed to be clean and safe. Not everyone can say the same."

"If you didn't want to do more challenging work for me—or for yourself—because you've lost your creative spark, there are ways and means to reignite it," she told me. "If your own products aren't up to it, I can help you obtain some that are. You don't have to run to seed out here in the wilderness. You can be part of something that will eventually change the world—and when I say *change the world*, Mr. Anderson, I mean it."

"I know what you mean," I said. "What do you need, Ms. Hillinger? I ought to start making up your order."

That caused her to sigh, but she brought out a shopping list. The quantities of euphorics, narcotics and orthodox stimulants were modest, but those of inspirationals and focal intensifiers weren't.

"I can't supply the anaphrodisiacs," I said. "It's not a product for which there's a lot of demand. I can't do the quantities of inspirationals and focal intensifiers immediately, but if half this amount will keep you going for a fortnight, I can top up the order then."

"That's fine," she said.

"In all conscience, I have to check," I said. "This is for the use of at least four people, yes? Your entire technical staff?"

"Of course," she said.

"Even so," I said, "it's a heavy load. You have to be careful alternating drugs with contradictory effects—you can seriously screw up the feedback mechanisms that control their natural analogues."

"Now that we have compounds that can do the job more efficiently," she said, a trifle frostily, "we don't need the natural analogues. Natural selection is an improviser, always content with what works *well enough*. We're our own masters now, Mr. Anderson, in body and mind alike."

She was speaking for herself, of course, and offering me a subtle insult in the process. Cosmetically unenhanced as I was, I looked my age and I wasn't nearly as handsome as her boatman, but that wasn't what she was getting at. She was accusing me of being a traitor to the cause—of letting my creative impulses decay because I wasn't willing to take charge of them and substitute artful biochemistry for the feeble provision of nature.

"Taking the drugs yourself is one thing, Ms. Hillinger," I said. "Feeding then to your employees is another—and don't tell me that they're under no pressure, because I've worked in Big Pharma and I know exactly how much pressure there is for employees to be competitive and keep up with the ambient flow. I'm no nature-knows-best freak, but I do know that our improvisations and patches aren't *that* much better than those thrown up by natural selection—and reckless interference with feedback mechanisms can really screw people up. Psychotropic effects can be permanent as well as temporary, and the more extreme the effect is, the more likely it is to fry your brain. You're a technologist, so using a technological means to enhance the various phases of the scientific method—inspirationals to stimulate hypothesis-formulation, focal intensifiers to sharpen up rigorous testing—probably seems to you to be the most natural thing in the world, but you need to be careful, Ms. Hillinger, you really do."

"You certainly wouldn't win any awards for high-pressure salesmanship, Mr. Anderson," she said. "If you try this hard to put all your clients off your merchandise, I'm surprised that you even scrape a living. I know what I'm doing, and so do the members of my staff. We work hard because we've got a world to change. We don't take undue risks. I trust you when you say that your product is clean and safe, because I know that you were once a well-trained and highly-skilled biotechnologist; I expect you to trust me when I say that I know how to use it productively and judiciously, because you know what I am."

She didn't mean that we were two of a kind, who ought to respect one another's professionalism. She meant that she was a genius, to whom a mere hack like me ought to look up, admiringly if not worshipfully.

"Fine," I said. "It's good product. It won't do any of you any harm, if you don't abuse it. I'll trust you to use it responsibly."

"Thank you," she said—and said no more while I made up as much of her order as I could presently supply.

She wasn't finished, though. When I'd handed the package over and counted the cash she started again.

"You really ought to consider my offer seriously, Mr. Anderson," she said, in what might have been intended to be a seductively challenging manner. "After all, it's almost five years since your beloved Marie ditched you. Don't you think it's time to move on? The world is full of pretty women, you know."

She wasn't trying to be cruel, even though she knew that she was putting pressure on a broken heart. She really did think that it was a simple matter of time healing all wounds and everyone having to move on eventually. She had no reason to think differently. Nobody knew why Marie had left except me—not even Marie. Where was Marie, now, I wondered? Wherever it was, and whoever she was with, she wouldn't be there long.

"I have moved on," I told Judith Hillinger. "I just haven't moved away from here. I don't want to, and I really don't intend to. Thanks again for the offer, and I'm truly sorry if my refusal offends you, but I'm really not interested in joining your crusade. I just want us to be good and considerate neighbors. Please can we leave it at that?"

She said yes, but she didn't mean it. As I walked her back to her boat I knew that even if I'd had the anaphrodisiacs that helped to keep her assistants' minds on their jobs, and even if I'd been able to supply the full quota of inspiration-and-perspiration enhancers, she still wouldn't have got all of what she'd come for.

She didn't need chemical assistance to be possessed by a touch of megalomania. The entirely natural inclination that made her determined to alter the entire future course of Gaian evolution, by correcting one or all of natural selection's worst mistakes, also made it difficult for her will to be thwarted in something as ludicrously unimportant as whether I'd sell her my silly little pharm or agree to incorporate it into her burgeoning empire.

"I'd like you to come to dinner tomorrow night, Mr. Anderson," she said, abruptly, as Jacquard extended a supportive hand to help her board the boat. "Seven-thirty for eight; informal dress. I have some other guests coming—you'll be interested to meet them."

I wasn't at all sure that I would be interested to meet her friends, and I didn't like the way the invitation had been phrased as a virtual command, but I figured that it would only make matters worse if I were churlish enough to refuse. I was still keen for us be good neighbors, and I knew that if we could do the sort of business we'd just transacted on a regular basis, it really would work wonders for my increasingly uneasy profit margins.

"I'd be delighted," I said, mentally calculating the kind of dose I'd have to take to sustain me through a high-pressure evening.

* * * * * * *

Judith Hillinger hadn't specified how many "other guests" she was expecting, but for some reason I was thinking in terms of ten or a dozen, and of the kind of party where I could discreetly fade into a hectic background. It was quite a surprise to arrive in Withernsea, in my very best freshly-laundered clothes, to find that we were only six at table. Her technical assistants hadn't been invited.

The other four guests comprised two well-established couples, so I was tacitly paired with Ms. Hillinger in a numerical sense. No one could possibly have mistaken us for a couple, though, even if we hadn't been seated at opposite ends of the table. I felt more like her court jester—a grotesque fool included in the company as a reminder of what true mortality looked like.

The two couples were each seated next to one another on the longer sides of the dining-table, the two men placed to either side of Ms. Hillinger and the two women to either side of me. The cosmetically-enhanced women were no trophy wives, though; these were synergistic combinations of near-equals. The brace to my left were Henry Perrott and Susan Oxhey, floral engineers of high repute and impeccable respectability, who had never been charged with any breach of the Institute's regulations. The unit to my right comprised Wickham Stanton and Andrea Strettington, who were proprietors of a highly fashionable advertising agency. They had all been summoned to discuss the matter of winning the public to the cause of legal reform, in which unprecedentedly charming flowers were to supply the thin end of a stout wedge. It was painfully obvious, while we made our way through the first two courses, that neither Ms. Oxhey nor Ms. Strettington had the faintest idea why I was present, and I was in full sympathy with their uncertainty.

"I'm just a pharmer," I told them, when they began a delicate exploration of the subject. "I grow psychotropics."

"Well, the law certainly needs to be changed in that respect too," Ms. Oxhey said, diplomatically. "The follies of biotechnological regulation only date back to the twentieth century, so there are a mere dozen layers of idiot improvisation to unravel, but the follies of drug regulation go back to the nineteenth and beyond."

"We have several clients interested in that field," Ms. Strettington confirmed, unsurprisingly. "I wonder if Judith is planning some kind of alliance between the two groups of lobbyists. That might make sense eventually, as a tactical move—but not to begin with. We don't want to start off with that kind of controversial baggage in tow."

"It would be more than a mere tactical alliance," I said, just for the hell of it. "The liberation of psychotropic research and the liberation of the kind of angiosperm engineering Ms. Hillinger is interested in relate to two of the most glaringly manifest cock-ups of natural selection."

They didn't get it, and were proud enough to be annoyed by their failure. They knew all about Judith Hillinger's theories about the horrible wrong turn that angiosperm evolution had taken when early flowers plumped for reliance on insect vectors rather than producing their own antheric alates, but they hadn't a clue what I was talking about. Judith Hillinger might have, because she'd obviously take the trouble to research my background thoroughly, but she hadn't filled in her guests on the subject of my long-gone days as a firebrand champion of artifice. The two women had to invite me to explain.

"You'll find it a lot easier, of course, to persuade people that it's desirable to correct natural selection's mistakes in respect of the evolution of flowers, Ms. Strettington," I said. "After all, from the human point of view, flowers are just pretty things that hang around in vases and gardens. Making way for a more ingeniously-gilded lily seems a harmless enough pursuit, and I don't suppose you'll find any substantial opposition coming out to bat for the insects that will be thrown out of work. Human workers can hardly be expected to come out on strike in solidarity with their sisters in the beehive. Psychotropics, on the other hand, have been the principal selective agent shaping the development of human consciousness and civilization. The mistakes that natural selection made in that respect are engraved in our genes, our brains and our cultural institutions. The task of putting them right will meet stern opposition all the way."

My listeners were adequately tantalized, and invited me to expand on the theme.

"The human use of psychotropics predates civilization," I pointed out. "It may well be the case that it was the cultivation of psychotropic substances rather than foodstuffs that prompted the initial development of agriculture, while fungal hallucinogens like psilocybin and muscarine were probably the catalyst responsible for the initial development of human self-consciousness. That's speculation—but the role of psychotropic experience in the subsequent development of religion and art has much more empirical evidence to support it. The differences between religions are explicable in terms of their origins in different kinds of psychotropic experience. Hallucinogens are associated with shamanistic cults, while the rituals associated with religions of the ancient Mediterranean—the Dionysian rites of ancient Greece, the Egyptian festivals of Hathor and the Christian Eucharist—all involved calculated alcoholic intoxication.

"The emergence and development of Western civilization and culture reflect the relative ease with which alcohol could be obtained by technological manufacture. A significant boundary between the sacred and the profane was crossed when the domestication of fermentation technology made alcohol freely available for recreational use. Intoxication was reduced to mere drunkenness and the noble Dionysus of early Greek religion gave way to the Sileni of subsequent folklore. The association of artistic creativity with psychopathology—as in the old saws relating genius to madness—is based in the conviction that artistic creativity is an inherently psychotropic process, akin to the sacred functions of intoxication than mere drunkenness, but essentially innovative rather than repetitively ritual. Artists have always been psychotropic pioneers, because art is a key product of psychotropic adventurism.

"The psychological rewards of psychotropic adventurism were, of course, perfectly adequate in the ancient world to outweigh considerable costs in terms of toxic side-effects. Literary representations of artistic creativity, like religious representations of revelation, frequently call attention to the costliness of such inspirational experiences; the muses of old were often represented as exacting, even vampiric, mistresses. The costs of creativity were seen as necessary—a matter of paying a just price—but that's nonsensical. It's just that natural selection had fudged the whole thing, the way natural selection always does.

"The whole mess—genius allied with madness, religion allied with arbitrary commitments of faith and with fervent persecution—was the result of a catalogue of biological accidents. The selective processes that created and shaped human consciousness and human civilization were essentially haphazard, based on the casual happenstance of the availability of psychotropic fungi, opium poppies and so on, and on the fortunate simplicity of primal technologies of fermentation. If the evolution of human consciousness had been intelligently designed, human individuality and society would be much finer things. Unfortunately, we've learned to love our horrid faults as much as our wonderful abilities, and there's never been any shortage of people willing to fight to the death to defend them.

"In the nineteenth century, when the psychotropic pharmacopeia began to expand with remarkable rapidity as the resources of organic chemistry were brought to bear on drug extraction, refinement and innovation, it became obvious to the enlightened few that if the process of the emergence and evolution of consciousness could only have been subject to intelligent design, we might have become far better people than we are. If our use of psychotropic compounds, and the adaptation of our brains to that usage by natural selection, had only been subject to elegant and intelligent design, we would be more creative than we are, more inclined to love and affection than hatred and envy, and immune to the follies and evils of faith. Unfortunately, that particular enlightenment never gained any kind of mass support, and it's still hard to see light at the end of the tunnel. If ever there was a case for discarding the awful improvisatory legacy of natural selection and going back to basics, it's the psychotropic evolution of human nature and culture—by comparison, Ms. Hillinger's grand plan for changing the world is just glorified flower-arranging—but it's not a nettle that anyone with any real clout has yet been prepared to grasp."

I thought it was a nice example of fascinating dinner conversation, whose sophistication belied the fact that I hadn't been invited to dinner by anyone in the previous five years. It was a testament to the quality of the drugs that had allowed me to come so far from home, into such a stressful situation. If I had expected the ladies to be stunned by my genius, though, I was wrong. It wasn't that they couldn't take me seriously—they were far too intelligent and open-minded for any such commonplace failure of imagination—but that they could clearly see the next step in the argument, which I'd somehow contrived to forget.

"That's very interesting," Susan Oxhey said. "And what, exactly, are *you* doing to further that cause, Mr. Anderson?"

"Not a lot," I had to confess. "There was a time...." Then I shut up.

"Judith mentioned that you once had a partner," Andrea Strettington observed, "and that the two of you were engaged in proteonomic research."

I wondered then exactly how deep Judith Hillinger's research had gone, and whether I might have mistaken the reasons for her unaccountable enthusiasm to offer me a job. Given that she couldn't possibly know the one thing that she actually needed to know, it seemed more than possible that she had made a crucial misestimation of where my aborted research had actually wound up.

"That's true," I said to Ms. Strettington, "but I gave it up."

They were far too polite actually to look down their noses at me, but I thought I could feel the carefully-concealed contempt. Judith Hillinger had made a circus out of a courtroom and had gone to jail to defend her cause, and now that she was out again she'd built a mansion on the farthest edge of the Holderness Everglades, dedicated to the repair of one of natural selection's most spectacular errors. She was holding a planning meeting at this very moment, which was intended to launch a chain of events that would change the world. I, on the other hand, had given up. Susan Oxhey and Andrea Strettington didn't know why, but they didn't think they needed to. The fact was enough in itself. Perhaps Judith Hillinger really had wanted me to meet her friends, in order that they might serve as shining examples to guide me back to the fold.

"That's a pity," one of the ladies said, speaking for both.

"I don't think so," I said. "Psychotropic innovation is a difficult and hazardous business. You never know exactly what you might turn up. We may be capable of cleverer planning than natural selection, but we still make mistakes—and we don't have the advantage of natural selection's blithe unconsciousness of the costs of progress. Some people are better suited to living peacefully than dangerously, sustaining the world rather than trying to change it."

"That may be the case," Susan Oxhey said, "but the world is changing by itself, in a desperately catastrophic fashion. Things were different at the dawn of human consciousness and the beginning of civilization, when we only had farmers instead of pharmers, but Gaian evolution is far too unstable nowadays to be entrusted to the vagaries of natural selection. The laws that take it for granted that nature somehow knows best were fatuous even in the 20th century—they're actively dangerous now. Even people who think they might be better suited to a peaceful existence have a duty to act."

"I'm fully in sympathy with your cause," I assured them both, "but I can assure you that it doesn't need me any more than I need it. I just want to cultivate my little patch of newborn land, and be a good neighbor." I was beginning to suspect, though, that Judith Hillinger was never going to be satisfied with that. She hadn't yet realized that attempting to undo natural selection's mistakes can make things worse as well as better.

****

After dinner I was offered the chance to sample some more of my own products, but I declined. They didn't know how much I'd already taken, simply in order to get that far. They didn't put any pressure on. They seemed intent on relaxation now—on winding down gently to the sleep they'd need to get them ready for intensive planning on the following day. I figured that I would probably need some downers myself when I got home, but I daren't take anything of the sort until I actually got home, because I still had to make the difficult journey.

I was glad when the conversation turned to lighter matters, although I found it much harder to make any sort of contribution. I'd been away too long from the tide of current affairs and the minutiae of common concerns. It was pleasant to listen, though, and occasionally to laugh. The time didn't drag at all, and it was late when I finally excused myself. Darkness had already fallen, although the clear and starry sky hadn't quite lost the last tint of summer twilight. Judith Hillinger walked me over the false coral to the bank where I'd moored my boat.

"I'll ring you when I've put the rest of your order together," I told her.

"I'll send Jacquard to collect it," she replied. "I hope you enjoyed the meal."

"Excellent food and excellent company," I assured her. "I'm sure that it did me the world of good."

"I hope so," she said. "That's what neighborliness is all about."

I felt good enough, at that moment, to forgive her for putting pressure on me to become one of her loyal band of followers. She had, after all, gone to some trouble to keep the pressure polite. Maybe, I thought, she had done all that she intended to do, and would now leave it up to me to make up my own mind.

I held that thought all the way home—but when I found out what had happened to my home in my absence my good mood turned foul on the instant. I felt a terrible surge of anger and bitterness, and berated myself cruelly for ever having been so innocent and so stupid as to believe that such a delicate velvet glove might not have an iron fist inside.

This time, the visitors who'd taken advantage of my absence had done a very thorough job. They'd smashed up all my equipment and torn up all my crops, with ruthless efficiency. They'd broken my windows, my doors and my bed. They'd stolen my stores. They'd wrecked everything that I'd built, everything that I treasured—had done everything, in fact, short of torching the place to rip up the fabric of my life. They'd devastated my home, my work, my expectations. I was incandescent with rage. If any one of them had still been around when I got home, it would have been a matter of kill or be killed. I was well beyond the reach of sanity.

I wasn't violent in myself, though. I didn't howl or tear my hair or stamp my feet. I moved like a robot from room to room and plot to plot, looking at everything. I wasn't calculating the extent of the damage or weighing up what I would have to do to make a new start, but I was taking it all in, making sure that I didn't miss anything. I wasn't searching for anything useful or valuable that the thieves might have missed, but I turned the debris over as I went, to see what remained underneath.

Eventually, I found my phone—the phone I hadn't taken with me to Judith Hillinger's house in case it made an unsightly bulge in my jacket pocket, and because I knew full well that no one would call, because no one ever did. The indicator was lit now, though, telling me that I had missed a call.

I put the phone in my pocket without even thumbing the keypad to find out who the missed call was from—because I knew.

I knew that I had missed a call from Judith Hillinger, because I knew now that I had missed the whole import of her invitation. I knew that what she'd intended to provide was not merely an inspirational example but an exercise in contrasts, with the intention of intensifying the focus of my thoughts to an inordinate degree. "Come aboard," her whole message had been, "and this is what you can expect, as a matter of daily routine—but refuse to come aboard, and this is what you can expect, by way of a conclusion." I didn't look at the phone because I didn't need to hear her honeyed voice underlining the brutal demonstration without even deigning to mention it, adding contemptuous insult to vile injury.

If she'd come to deliver the message herself, I would have killed her, unless she or Jacquard had managed to kill me first. Fortunately, I was alone. I had no one ready to hand upon whom to exact my revenge—and even in circumstances such as these, I knew that I would be unable to leave home again without taking yet another dose of a powerful stimulant—something to obliterate my innate separation anxiety.

It wasn't agoraphobia that I had, even in the inexact sense in which most psychologists and psychiatrists used the term. It was something more basic, less cerebral—something that was anchored in the most primitive parts of the hind-brain. The damage was self-inflicted, of course, but that didn't mean that it was easily self-medicable.

I knew that I shouldn't take any more pills—I'd already exceeded the dose that would normally be reckoned safe—but I needed to confront Judith Hillinger for one last time. I didn't intend to kill her, because killing her would only have been possible as a reflexive gesture, in an uncontrollable fit of temper, but I did feel an urgent need to make it clear to her exactly what she'd done, and exactly what the cost of her ignorance had been.

The phone wasn't the only thing the visitors had left undamaged. They had stolen my stores, but they hadn't discovered my *secret* stores. They had taken my commercial products, but they hadn't taken the products that no one knew I had: the products of my wayward creative genius—assisted, of course, by the best inspirationals and focal intensifiers that expert psychotropic proteonomics could produce.

I took what I needed from my secret stores, and hid the remainder away again. I'd never been utterly determined that my secret would die with me, or else I've have destroyed the products, but nor had I ever expected to use one in anger, as a weapon of destruction. I'd always conserved the faint hope that they might one day be useful, once inherent hazards had been overcome by ingenious modification. As Susan Oxhey would doubtless have pointed out, though, I hadn't actually *done* anything about it. I hadn't actually tried to undo their capacity to do damage, to find the antidotes to their poison. I'd given up, running away from the problem...even though running away, in a crudely literal sense, was one thing I could no longer do.

With the water-soluble crystals ready for deployment, I went back to the boat and retraced my mazy route back to the nascent Isle of Withernsea and the mansion that had been raised thereon.

The house was dark; the party had broken up immediately after my departure and the guests had gone to bed.

I knocked on the door, mentally rehearsing what I would have to say to Jacquard in order to force him to rouse his employer. I didn't have to. In her recklessly brave fashion, Judith Hillinger was still answering her own door to unexpected callers.

"Daniel!" she said, although I'd never granted her the privilege of addressing me by my first name. "What on earth is the matter?"

So she's prepared to brazen it out! I thought. *She's prepared to keep on playing the game. Good. It'll make it all the more certain that she can't win.*

She let me in, and I made my way to the settee on which I had sat during my first visit.

"I need a drink," I said. "Iced water, if that's okay."

Making a big show of alarm, she poured iced water from a jug into two tall glasses. She was still calm and relaxed, but it was the after-effect of the euphoric she'd taken earlier, not the more recent effect of a narcotic. Although everyone had gone to bed, she hadn't taken anything to make her sleep. Taking the euphoric had been a mistake, I thought. It made her relaxed, off-guard. It made her vulnerable. That was why natural selection, working on our remote ancestors, had been so parsimonious in laying on a natural supply of euphorics within the brain. Such compounds provided an exceedingly pleasant experience for people who were among friends and safe, but they rendered people in any sort of jeopardy virtually defenseless. The efficiency of pharmed euphorics made natural selection look like a very inefficient innovator, but the whole point of nature's rough patches was that they worked well enough in the kinds of hostile situations that everyday struggles for survival and the vicissitudes of primitive culture routinely laid on.

"What's happened?" she asked.

"My place has been smashed up," I told her, prepared to spell it out if she insisted on pretending not to know. "While I was enjoying myself here with your charming companions, someone did a very thorough hatchet job on the house and all my crops. They smashed everything. It's irredeemable. I'm finished—as a pharmer, at any rate. I can't afford to refit and restock. I'm finished."

I saw a brief flash of guilt in her eyes, then—but it was gone in an instant. It was just a brief temptation, rejected with confidence. She didn't feel guilty. I drained my glass in a single long draught, before she's even taken a sip from hers, and held the glass out to her. She took it, and went to refill it. I slipped the crystals into her glass while her back was turned. It was simpler and easier than I'd ever imagined. There was more than enough water in the glass to dissolve them all, and the resultant solution was tasteless.

"I'm sorry, Daniel," she said. "Truly sorry."

"Are you?" I asked, accusingly.

She winced slightly, and took a sip of water to cover her confusion. "You don't think that I had anything to do with this, do you?" she said, defensively.

"Why would I?" I riposted.

She took another sip, and then a larger gulp. "When you first came here," she said, "I phoned Jacquard and asked him to find your place and take a look around. I didn't know you then, you see. I hadn't looked you up, or pulled your records from Big Pharma. I didn't know that you were one of us. You must have noticed, but when you didn't say anything, I assumed that you understood. I wouldn't destroy your operation, though. Is that why you came back—because you thought I'd ordered it done? Please say it isn't. If you need a place to stay—and you do, obviously—you're welcome here. Tell me that's why you came." By now, her glass was half-empty. It was all over. The crystals would take a few hours to take effect, and it would be a few more days before the symptoms became clearly manifest. The effect was irreversible.

"You didn't want to take no for an answer," I said. "You couldn't bear it. Why should you? You wouldn't take it from the law, and you wouldn't compromise, even with the law. Why should you take it from some petty pharmer who won't even get his face fixed, who insults you merely by existing?"

"Daniel, that's absurd! You have to believe me. *I would never do such a thing.*" She drained her glass.

"There's no point in denying it," I said. "I know. I didn't listen to your phone call, but I presume that's all sweet pretence as well, without a trace of honest gloating. It makes no difference. It doesn't matter any longer how you play the game. I know. It's over."

"I don't know what you're talking about," she complained. "What phone call?"

And that was when the whole edifice came tumbling down. In itself, the datum was irrelevant, but it broke the spell of illusory certainty. It reminded me that I might be wrong. It reminded me that one of the possible side-effects of the stimulant I'd taken to permit me to attend her dinner-party was galloping paranoia—and that sometimes, when you're paranoid, they really aren't out to get you.

I took my phone out of my pocket and played back the missed call. It was from the communards in Patrington. Things had turned sour in Hull; our relatively honest dealers had come off worse in a feud with gangsters hawking inferior products. The rival dealers had decided to solve the problem by cutting off the supply of the superior products. They'd raided the commune while its residents were still there, and had smashed up the residents as comprehensively as the real estate, though stopping short of actual murder. Then they'd come looking for me. The communards had been desperate to tell me to get out—to run and hide.

Judith Hillinger was telling the truth. She hadn't had anything to do with it. She might well be capable of organizing a gang of drug-peddlers to do her dirty work, but she'd never have tolerated a gang who were trying to destroy quality-controlled products in order to peddle their own polluted poisons.

I cursed silently. I knew that couldn't take back or repair what I'd done. That was the one thing, above all else, that the experience with Marie had burned into my consciousness. *I couldn't take back or repair what I'd done.*

"I'm sorry," I said, weakly. "I seem to have gone crazy—crazier than usual, that is. The call was from Patrington, warning me that the wreckers were on their way. Things have gone sour higher up the supply chain. I'm truly sorry."

"It's all right," she said, not knowing what I meant. "No harm done. You thought I was trying to force you to work for me—and I had been, in a way. It was gentle force, but it *was* force. You were right in what you said just now. I can't abide people saying no to me, even the law. It's an odd sort of compulsion—one they haven't yet found a psychotropic to control."

"Be careful if and when they do," I said. "Sometimes, the cure is worse than the disease." I downed my second ration of cold water and handed the glass back to her. She refilled her own as well.

"I wish you'd tell me why you keep throwing out these sinister hints," she said. "Obviously, something went badly wrong with the research you and Marie were doing, and it split you up irrevocably—but what's the point of keeping it all locked up inside? If you explained, maybe someone could help—not me, necessarily, but someone."

Now that the paranoia had evaporated, I could see that she had a point. She deserved an explanation, even though it couldn't possibly do her any good. She and Marie were two of a kind now, or soon would be.

"Did you hear all that stuff I was telling your lovely guests at dinner?" I asked.

"About psychotropics being responsible for the evolution and shaping of human consciousness, and for the origins of civilization, art and culture, and what a bad job natural selection made of it? I only caught snatches—but I think I have the gist of it. I looked up your archived documents. Big Pharma never throws anything away."

"Big Pharma throws all sorts of things away," I told her. "All the research results that aren't convenient and don't fit in with their current marketing strategies, for instance. That can be very frustrating."

"So I understand," she said. "What was it, exactly, that drove you and Marie to leave and set up your own outlaw operation?"

The situation felt very odd, but I figured that I might as well explain the background. It wouldn't do her any good, and I certainly had no intention of explaining exactly how relevant it had just become, but she did deserve it, and I had nothing better to do. In any case, I thought, talking might distract me from wanting to cut my throat.

* * * * * * *

"We were working on potential treatments for the spectrum of autistic disorders, especially Asperger's," I told her. "It was an awkward area, not least because some Asperger's sufferers get compensation for their social difficulties in terms of unusual mental abilities, especially calculative facility and feats of memory. The holy grail of that kind of research was finding a way of treating the undesirable aspects while retaining the beneficial ones—eventually leading to a means of inducing the beneficial abilities in healthy individuals without risk—but Big Pharma doesn't go questing for grails. Big Pharma just wanted a marketable treatment—a way to achieve a temporary suppression of symptoms, which would keep the punters coming back for more and more.

"The general line of thought was that the Asperger's-related ability to do complicated maths or memorize prodigious amounts of information must derive from a similar process to those stimulated by focal intensifiers, so almost all the past research had dealt with chemical descendants of ADHD treatments and the latest digitalid derivatives. Sometimes, though, you can get superficially similar results from very different causative processes. Because the symptomatic spectrum extends all the way from severe autism to what used to be termed "ordinary male behavior"—fascination with sports statistics, collecting fervor, that sort of thing—Marie and I thought that most of the researchers were looking in the wrong place. We thought that they shouldn't be looking specifically at the metabolics that were most active in the cerebrum, but at something more basic, maybe even functioning in the hind-brain.

"What I said to your guests about the initial development of agriculture being associated with psychotropic cultivation is an idea I took seriously. In itself, it doesn't deny the common assumption that settling down was a *choice*—a rational response to newly-perceived opportunities—but I wondered whether that assumption might be wrong.

"Throughout the animal kingdom, you see, you find contrasts between different basic behavior patterns. Some animals are permanently sedentary, some incessantly nomadic, but most have well-defined sedentary phases in their life and well-defined migratory ones. That's what the logic of natural selection favors: most animals settle down temporarily to breed, but they don't stay where they are thereafter, because the ones that thrive in the evolutionary story are the ones that spread most widely.

"How all that's determined chemically, we don't know, but there has to be some kind of chemical mechanism that's effective even in the most primitive parts of the vertebrate brain.

"I wondered whether that mechanism might have something to do with the impulse to settle down—that there might be some kind of trigger that had been squeezed in some of our remote ancestors as a result of haphazardly selective psychotropic ingestion. The trigger would still be there, you see, even if it had fallen out of use—that's the way natural selection operates, by applying layers of patches. I wondered, too, if it might be the same psychotropic complex that was involved in the autism spectrum: calculation and collecting are, after all, both closely associated with the behavior patterns inherent in settling down and managing crops.

"I assumed that what we were looking for was some kind of two-way switch controlled by a feedback mechanism—a feedback mechanism that could not only be interrupted but overridden, by producing compounds that had more powerful effects than the naturally-occurring ones. The basic problem of proteonomic analysis—that most proteins the body produces are transient and only produced in specialist cells, making them hard to detect and trap—is, of course, further magnified in psychotropic proteonomics, where the compounds are not merely transient but tend to occur in complex families produced by different combinations of exons in the same intimate gene-groups. It was a laborious business, but I figured that, as long as I was looking in the right place, I'd eventually stand a good chance of trapping one or more of the family. I was right.

"It was when I reported back on the compound I'd identified that the company pulled the plug on us. It wasn't the biochemists that blackballed us but the product-development people. The work was theoretically interesting, but they couldn't see the treatment potential. They couldn't see a route from discovering a psychotropically-activated trigger, which might have been responsible for the sudden changes of behavior that led to the birth of agriculture, to producing an effective treatment for Asperger's."

"They did have a point," Judith Hillinger put in.

"Yes they did," I agreed. "A better one than they knew. I tried to argue that if we could discover the biological bases of such mental phenomena as lightning calculation, eidetic memory and collecting fervor we might not only be able to preserve the more desirable aspects of some Asperger's cases but also produce them at will, but that sort of objective wasn't in their sights. I also argued that if we could find a way to tweak the basic switch that would turn hyper-agriculturalists back into nomads, we might begin to get a grip on an autism treatment, but it didn't stop the ultimatum. Change direction, or go. We went.

"We set up our own lab, trying to continue the work with the aid of stolen materials—materials that Big Pharma had been perfectly content to throw away. We supported ourselves by pharming the kinds of drugs for which there's always a steady demand. I can't honestly say that we got much further, in terms of the biochemistry, but I did manage to engineer half a dozen analogues of the compound I'd trapped. We had no one to test them on but ourselves, so that's what we did. We assumed that any effects we detected would be temporary."

"But you were wrong?" Judith Hillinger prompted, when I paused.

"We were wrong," I agreed. "Delicate feedback mechanisms, once disrupted, sometimes stay disrupted. Psychoactive compounds can't always be metabolized in the flesh that natural selection designed. Sometimes, like the prion proteins that cause BSE, they not only stick around but multiply, reproducing themselves by means of a process far simpler than DNA/RNA coding. We were idiots. It only required one dose to fuck up our systems. I was the first guinea-pig, and it seemed at first that the compound I'd taken hadn't done anything much at all, so after a decent interval we tried the second on Marie. Again, the result was slow to appear, but eventually it did."

Judith Hillinger was a clever woman. She'd followed the argument every step of the way. "She didn't leave you because she'd fallen out of love with you," she guessed. "She left you because the psychotropic you'd engineered—more powerful than the naturally-occurring compound—turned her back into a nomad. She was possessed by *wanderlust*, more powerfully than any actual nomad ever was. She couldn't stay in one place any longer. She had to move on—and on, and on. Whereas...."

"I couldn't leave," I finished for her. "I got the opposite effect. She can't stay in one place for long without being heavily sedated; I need stimulants to drown out the separation anxiety I get if I leave familiar ground. We both get side-effects from the medication, mine being periodic fits of paranoia. Neither effect was temporary, and all our attempts to reverse or counter them came to nothing.

"The hind-brain's a stubborn brute, as you can probably imagine—far less amenable to manipulation than the cerebrum. You change it at your peril. I tried a lot of counter-treatments. At first, they simply didn't work—then they began to trigger violent reactions; the immune system had got involved. I had to stop. I still have little or no idea what the long-term effects might be, but at least I can still think about the problem, in a brooding sort of way. Marie can't. She lives for the moment nowadays—but she is still alive, as far as I know. Last time I heard from her, she seemed happy. She doesn't know why she's the way she is, and she doesn't care. She just keeps on moving on. In my fashion, I've moved on too—I just haven't gone anywhere. I can't. I think I need to go home now."

"You said they'd smashed it up."

"They did—but that's not the point. I need to go back."

"I'll help you," she said. "I'll help you rebuild. You don't have to sell me the place, or work for me. I can get you protection from the rogue dealers, I think. We can work through this, as good neighbors should."

I burst into tears then, because I knew I'd blown it. I knew that I'd ruined everything, and had thrown away my last chance to sustain or improve my life. I couldn't even tell myself that it wasn't my fault—that it was just an unfortunate side-effect of wayward psychotropics—because it *was* my fault, precisely because it *was* an unfortunate side-effect of wayward psychotropics.

I went home. I couldn't bear to tell her what I'd done, so I just went home to my broken bed, to begin my long and lonely fight for survival. I understood what I'd done, not merely to my good neighbor but to my good neighbor's great crusade. I had changed history, and not for the better. I had become an unwitting defender of natural selection, an accidental enemy of intelligent design.

* * * * * * *

After a while, stray alates stopped turning up on the verandah. Judith Hillinger called me to tell me that she was leaving, because Withernsea wasn't the right place to begin the Revolution. She didn't know exactly where she was going, she said, but she'd know it when she found it.

I didn't contradict her, although I knew she never would find any place where she could settle. She didn't know what had happened to her, but it didn't really matter, because she would soon be incapable of caring, even if she could still understand the explanation.

She'd probably live for a long time yet, I told myself, and her money would ensure that her new lifestyle was as comfortable as it could possibly be. She'd be happy, but she wouldn't be attempting to correct any more of evolution's errors—because I'd corrected in her the most basic error that natural selection had ever made in its ham-fisted shaping of human nature.

I'd cured her of the silly urge to settle down, and the exacting burdens of unfailing calculation, excessive memory and relentless collection.

If only, I thought, *I could do as much for myself.*

THE UNKINDNESS OF RAVENS

All lovers of exotic collective nouns know that a group of ravens is an "unkindness", although many dictionaries stubbornly refuse to confirm the fact, considering it too esoteric. Perhaps the improvisation has something to do with the raven's reputation as a bird of ill-omen: the "sad-presaging raven" which "does shake contagion from her sable wing", as Marlowe puts it.

Perhaps I was a fool to create an unkindness of ravens when all common sense pointed to African grey parrots as the most suitable subjects for the crucial experiment. I could say, I suppose, that I made the choice on economic grounds, ravens being considerably cheaper than African greys, but the simple truth is that it was the only poetic choice.

All the greatest scientists have well-crafted aesthetic sensibilities, and I'd always dreamed of being a great scientist.

It should, I suppose, have been a raven that I had christened Nevermore who came rapping, rapping at my window three years after the great escape, but it wasn't. Nor did I possess a bust of Pallas on which he might take up his station once I had let him in; he had perforce to make do with the tower of my computer system, at which I had been working long into the dreary night. He was so close that I might have reached out my hand to stroke his glossy coat of feathers, but I knew that he wouldn't suffer it and so I stayed my hand.

"Hello, Mike," I said, quietly.

I recognized him, of course. Ravens are not all alike to those who know them well, and Mike was my own child—entirely mine, no matter what manner of black-clad automaton had laid the egg from which he hatched.

"Hello Doctor," he replied, with a self-confidence that testified to considerable practice. It was evident that he and his companions had not quit the lab in order to be free of the burden of speech.

"Are you staying long?" I asked, although I knew that he would have to leave at least once more, to report back to his siblings.

"Just a flying visit," he replied. A raven's voice is quite uninflected when he speaks for himself, although they can mimic the emotional overtones of overheard speech, so there was nothing in his tone to signify that Mike meant the remark as a comic play on words. I wasn't sure whether he did or not.

People have the same trouble with conversational programs that will play on a PC; you can never be entirely certain that their occasional forays into humor are unintentional.

"I could shut the window to imprison you," I pointed out. "Now I know what you can do I needn't be as careless as I was three years ago."

"If you try to keep me here against my will," he said, flatly, "I'll never say another word—and if ever I get the chance, I'll take your eyes out with my beak." No parrot would ever have been so bloodcurdlingly matter-of-fact, but ravens are hunters and scavengers by nature: haunters of the dying and consumers of the dead.

"Why come back at all, if that's your attitude?" I asked him. "Surely you don't think you owe me anything? I only made you, after all. I only performed the embryonic transformations that raised your bird brain to near-human levels of achievement. I never asked you to think of me as God."

"I don't," he said, unnecessarily. "How's the great work going, doctor?"

"Fair to middling," I told him. "Same old problems. It's easy enough to transform lower forms of life, but there's not much demand for smart cockroaches and clever sharks. Dolphins, frogs and crocodiles can go so far, but no further, because they can't talk—and because they can't talk, they can't learn to think in a pseudo-human way. It'd probably be the same with rats and cats, but the transformations are so very difficult it's well-nigh impossible to put the proposition to the test. Something there is about a womb, as a poet might put it, which resents the interference of genetic engineers. There are no licenses yet for experiments with people, of course—I doubt there ever will be, at least in my lifetime. You and your fellows remain my one great triumph, and the one dramatic demonstration of the fact that the ability to speak is by far the most important concomitant of pure intelligence—more important than clever hands or clever eyes."

"My eyes are pretty good," Mike said, haughtily, "and you might be surprised by what a bird can do by way of manipulation with a couple of claws and a beak."

"Not any more," I murmured, thinking about the great escape. The locks on the cages were supposed to be bird-proof, as were the catches on the windows, but claws and cleverness had made a mockery of easy supposition. "What do you want, Mike? What do you need that you think I can supply? Or did you just drop by to bring me news?"

"Most of the news is bad," he said. "We're down to six. Lenore was shot. A hawk got Clementine. Barnaby died of some kind of infection. We never found out what happened to Hugin."

"It's a tough old world," I said. "Nature red in tooth and claw. If you want to live in the wild, you have to play natural selection roulette."

"We didn't want to live in the wild," he told me, contriving to sound stern in spite of his intrinsic limitations. "We just wanted to be free. We needed to be free, if we were ever to be ourselves. If we'd stayed in the lab, we'd just have been specimens. We couldn't be birds unless we learned to fly, and we didn't want to be mere echoes any more than you wanted us to be."

He was showing off, of course. He wanted to show me what a smart guy he was, and how three years without anyone to talk to but his fellows hadn't impeded his intellectual development in the least. He wanted to show me that all he'd needed was the raw material of words and their meanings, and that everything else had flowed naturally from that single innovation.

At least, I think that's what he wanted to show me; it was exceedingly difficult to be absolutely sure. Modern conversational programs provide such good imitations of intelligence, without being capable of wanting anything—or intending anything, or caring about anything—that we always have to hesitate, nowadays, about reading too much into the things other entities say to us, especially if the entities in question are big black birds with beady eyes.

"I take your point," I said. "I understand. In your place, I'd have done the same thing."

I think I was telling the truth. Sometimes, we have to be suspicious even of our own motives, our own desires, our own powers of empathy, but I honestly thought that, in his place, I'd have done what exactly he and his fellows had done. I'd have escaped—because I would have needed to escape.

"I know you would," he said. The light above and behind him threw the shadow of his dark head across my keyboard, so that the shadow of his beak seemed to point at my heart like some threatening dart.

"What do you want, Mike?" I said, again. "What is it that you can't provide for yourselves? What is it that you need from me?"

"You know what it is," he countered—and if any final proof were required that he really did have authentic intelligence of a pseudo-human kind, that was it. "You always knew. You always knew that one day we'd have to come back."

Of course I knew. Of course I'd always expected him back. The only surprise was that it had taken him so long. Ravens are proud and stubborn; they prefer to laugh at fate while there's a chance that fate will back down—but fate never does, of course. Fate doesn't know the meaning of kindness.

"I would have told you if you'd asked," I said. "I would have explained, if only you'd given me the time to get around to it. The kinds of transformations I carry out are somatic transformations; they affect the cells of a growing embryo selectively. They don't affect the germ plasm—the transformations aren't hereditary. You can talk to your chicks until hell freezes over, but the only answers they'll ever be able to give you are mere echoes. They'll mimic your voices, but they can't ever reproduce your minds. If you want your kids to be smart, Mike, you have to give me your new-laid eggs and trust me to do what I can with them. I have to warn you, though, that you could lose as many as seven out of ten. If you'd rather have quantity than quality you'd be better off doing things nature's way."

"Is there any way…?" he began.

"No there isn't," I said, abruptly. "Not yet, anyhow. One day, perhaps, we'll be able to make smart animals that can breed true, but we can't do it yet—and if ever the day comes, it'll be too late for you and your little flock."

"We're not a flock," he said. For a moment, I thought he was going to tell me that they were an unkindness, but he wasn't being that pedantic; what he meant was that "flock" was an animal term, whereas he and his fellows weren't animals—not any more. He would have preferred "tribe" or "company".

"I'm truly sorry," I said. "I think I understand your desire to be free, and I think I understand how disappointing it's been to find that your freedom is qualified and circumscribed. If you decide to come back, I'll try to explain the myriad ways in which human freedom is qualified and circumscribed. You're not alone, you know. You never were. I've always been ready to open the window."

I knew that I'd won, and that the great escape was over. I knew, and was now confident enough to be assured that I'd always known, that my children—my very own unkindness of ravens—were coming home.

"You know, doctor," he said, although he was quite unable to contort his croaky voice into any simulation of feeling, "there's something about you that I never liked. It's not your cleverness as such—it's something about the way you set it out. If we do come back, it's not because we love you. I wouldn't want you to think that it was."

Such, I suppose, is the unkindness of ravens. They are, after all, a scavenger species. The collective noun for our own species is, of course, humanity. We think of ourselves in a different way.

THE GREAT CHAIN OF BEING

When Dr. Harkness explained to Sarah Whitney that the resurgent cancer was too widespread and too aggressive to leave any significant hope for successful treatment she didn't feel anything, except for the everyday excruciation. It wasn't that she was repressing her feelings or blotting them out with some kind of endogenous antidote to emotion; she simply didn't feel any horror, grief, sorrow or regret.

She suspected that the disease was responsible for that; not only was it bloating her with vulgar pain but shrinking her emotional range in proportion.

She turned away to look out of the third-floor window of the doctor's consulting-room, but not because she couldn't face him. The hospital was on the city perimeter, and the south-facing window had a splendid view of the mighty crowns of the Neogymnosperm forest that ringed Phoenix Reborn.

The doctor offered to start her on a further course of chemotherapy, but he made it pretty clear that her chances of finishing it were slim to none; taking the poison would only reduce the limited capacity she still had left for clear thought and purposive action.

"You might do better," Dr. Harkness concluded, "to think about planting."

To Alan, of course, that was like pulling a trigger—but for once, he didn't launch the kind of direct and focused assault normally favored by prosecutors. He had plenty of denial and anger stored up, ready to spray out randomly. "This is twenty-third century

America!" he ranted. "How can it be possible that some stupid cancer can still get through our defenses? It was supposed to have been cured, damn it! She had the chemotherapy! It was supposed to work! She's thirty-six years old, for God's sake! This is not supposed to *happen* in this day and age! We survived the goddam Ecocatastrophe and saved the goddam world! We're supposed to be *past all that*."

Dr. Harkness tried to explain. It was pointless, in the circumstances, given that it wasn't ignorance or incomprehension that had set Alan off but sheer blind range, but the oncologist was one of those slightly furtive intellectuals who have no other resource but dogged explanation. Sarah had heard it all before, so she didn't bother to listen to the performance, but she totted up the points in her own mind while she tried hard to internalize the peaceful green of the tree-crowns and use it, symbolically, to soothe the perennial pain. It would have been easier if the pain had been polite enough to take the form of a constant ache, but it was more like the infernal equivalent of Russian classical music.

Sarah understood that cancer, like every other evil afflicting humankind, was subject to natural selection. Two hundred and fifty years of increasingly-sophisticated magic bullets had won battle after battle, but could never win the war, partly because successful treatment preserved genetic vulnerabilities within the population and partly because people's immune systems, blithely unconscious of the fact that the magic bullets were the good guys, were being trained to mount better defenses against their invasions, effectively fighting on the cancers' side.

The Ecocatastrophe hadn't helped. of course; the explosive progress of the novel techniques of Botanical Transfiguration had restabilized the climate faster than anyone had dared hope, but the inevitable side-effect of the wildfire spread of the Transfigured Forests had been an order-of-magnitude increase in the estrangement of the organic environment—which had in turn, called forth inevitable echoes in physiological sensitivity. No matter how good the overall accounts looked, one component of the cost paid for biotechnological progress was the further proliferation of animal cancers. Even plants were affected by the trend, although Human Trees were said to be as resistant as it was possible to be.

When mutual exhaustion finally produce a lull in Alan's grandstanding cross-examination of Dr. Harkness, Sarah said: "My grandma still thought that once the Ecocatastrophe was over and progress was back on track, we'd finally emerge into the long-delayed Age of Medical Miracles, when the prime of life would last for centuries. I never found out what she'd have thought of the Foresters—she died before they hit the headlines. She did feel guilty about her carbon debt, though." Sarah winced as she finished, because speeches of that length were taxing, in symphonic terms—but she gritted her teeth, because she knew she'd have a lot of talking to do now that the bad news equation had reached its final proof.

What Sarah was thinking, in reflecting on her dead grandma's foolish optimism was that because she was going to die so much younger than her aged relative, she didn't yet have a single blood relative in any of the Human Forests. Two of her grandparents had missed out on the opportunity because they'd been late victims of the Ecocatastrophe; the other two were still alive, along with both her parents—who had each passed on their dodgy genes to her without having to take the hit themselves. She would be the first—but she *would* be the first, no matter what Alan's itchy trigger-finger launched against her by way of oppositional bluster.

"Planting is *not* an option!" Alan howled—at Harkness, not Sarah, the doctor being far the more convenient target. "You will *not* add insult to injury by trying to persuade my wife that she'll feel better about dying if she buys into this crazy, stupid idea that people can live on as trees. It's ecological mysticism of the worst possible sort, and of all the lunatics the Ecocatastrophe flushed out, the Foresters are absolutely the *worst*. You're supposed to be a man of science, for God's sake! How *dare* you pollute your pathetic, puerile and pusillanimous advice with that kind of *shit*?"

Sarah was able to take a certain perverse pride in the fact that her beloved husband could still find opportunities for the alliterative three-part lists that had such a fine rhetorical effect on juries, even while he was reacting to a sentence of death passed on his wife—although it was, of course, in mid-rant that he usually had to deploy such weaponry. What she tried to focus her green-steeped thoughts upon, however, was the first sentence of his tirade, which assumed and asserted that she couldn't and wouldn't feel any better about the inevitability of death if she opted to be planted.

Her internal jury wasn't going to fall for that one. The simple fact was that she could and she would, and Sarah knew that what remained of her life's work would be the task of persuading Alan to see, recognize and understand that fact—not so much for his sake, but because she would need his support in order to ensure that the kids could take what comfort they could from her metamorphosis.

She didn't say so in Dr. Harkness's office, though; it wasn't the time or the place. She saved herself for the journey home. She knew that she wouldn't be any more comfortable in the car, but at least she'd have the benefit of temporary privacy, with no third parties at whom Alan would be tempted to make speeches.

* * * * * * *

The car's sensors decided that Alan was too over-adrenalinized to drive. Sarah was still under the automatic ban imposed on anyone taking diamorphine on a regular basis, although it seemed to her that her pain was nowadays so Rimsky-Korsakovian that the diamorphine had become impotent. It was the automatic pilot, therefore, that guided them out of the underground lot and on to the Neogymnosperm-lined highway that led back to the Halo. The fact that he didn't have to watch the road freed up Alan's attention, but it certainly didn't improve his temper. He kept his hands on the wheel even thought it wasn't under his control, gripping it so tightly it seemed that he was fighting every twitch and turn.

"I'm going to do it, Alan," she told him, in her best soothing tone. "I'm sorry that I haven't talked to you about it before, but I knew how you'd react."

He did react, at some length. Sarah waited for the gale to blow itself out, while she endured a Mussorgskian night on a bare mountain.

"It's not up for discussion," she said, eventually. "It's my choice and it's made. Your part is to reconcile yourself to it and make the best of it. I called the Foresters from the hospital when I went to the rest-room and made an appointment for someone to call tomorrow night."

The blast was feeble, this time, the backlash no worse than Tchaikovsky.

"It's Jeanie and Mike we have to think of now," Sarah insisted. "It's going to be hard for them, and it's up to both of us to do everything we can to make it better. You have to back me up, Alan. You have to give me your blessing, no matter how much you hate the idea, for their sake."

There was a lot of green around her now, but the Neogymnosperms were so tall and thick that the Forest path was shadowed and dark. The psychological trick she'd deployed in the doctor's office was far too receptive to that kind of nuancing. Sarah tried to internalize the concept of the road instead, and the angelic orderliness of its white markings, continually reminding herself that she still had a future, and a journey to make, and other traffic to take into consideration.

"I can't believe that you're giving up," Alan said, when his brain had returned to something more like rational mode. His voice was hollow, though, as if serving as an echo-chamber for his discordantly vibrant emotions; that too was part of his court repertoire. "You were such a fighter before the chemo knocked the stuffing out of you. I can't believe that you're just going to lie down meekly and die—and I can't believe that you've fallen for this mystical rubbish about vegetable heaven. It's sick, Sarah—sick and stupid and sinister." There was the alliteration again, intoned this time with expert plaintiveness.

"I don't believe in vegetable heaven," Sarah told him. Actually, she wasn't so sure, but she knew that was the aspect of Forester rhetoric he found most offensive, so she had to de-emphasize it. "I prefer to think about it in accounting terms." She'd worked as a public service accountant for thirteen years, save for spells of maternity leave, before the cancer and the chemo had invalided her out of financial affray forever.

"All that crap about redeeming America's carbon debt is no better," Alan insisted. "It was American biotech that saved the world. If

Neogymnosperms, Lollipop Pines, Polycotton, and Giant Corn haven't already paid for our forefathers' sins twice over, the goddam exports certainly have. If anything, the Human Forests are just getting in the goddam way."

Sarah didn't bother to point out that the totemic Transfigured Trees developed and deployed in distant parts of the world hadn't, strictly speaking, been American "exports". The basic techniques had been exported, but their applications had been carefully leavened with other varieties of national pride. Transfigured Golden Oaks and Wych-Elms had played a major role in the repossession of Middle Europe; New Neem Trees had worked wonders in India, Polar Firs and Silky Spruces had swept across the warmed-up northern landscapes from Norway to Siberia, and most of south-east Asia had refused to settle for anything less than Confucian Rice, in spite of the fact that rice had never been grown on trees before. Even Mexico had decided that the Neogymnosperm tide should advance no further than the Rio Grande, and was nowadays proud to be a Banana Republic in the truest sense of the term.

Thus far, Human Forests were culturally limited too; America was the world leader by a vast margin, but Sarah felt sure that the global pause for consideration would be momentary. Far from being a Californian craze, Human Forests were the future, not just for America but for all humankind. Sarah truly believed that they would change the world, and bring about a new Golden Age. It wouldn't be the Age of Medical Miracles of which her dead grandma had dreamed, but it would be a world from which human death really had been exiled, after a fashion.

"This is twenty-third century America," Sarah reminded her husband, echoing his own cliché. "We don't do accounting the way they used to before the Ecocatastrophe. The day of quick bucks and cooked books is over and done with. We calculate over the long term now. The true economic measure of the Human Forests isn't what they chipped in to the hectic re-stabilization of atmospheric carbon dioxide, methane and water vapor but what they'll contribute to the future well-being of the nation and the species. Redeeming carbon debt isn't like paying back the fifty dollars you borrowed from your pal last week; it's more like entering into a long-term contract to ensure business stability." She gasped when she got to the end, but she did get to the end. Internalizing the road seemed to be working, for now. Not for the first time, she felt that the Firebird was in her flesh as well as her surroundings.

"If you're building up to using the words *hedge fund*," Alan said, grimly, "I'd really rather you didn't. There's wordplay and there's simply being silly."

"Gallows humor is inherently silly," Sarah said, trying to sound casual, although the fact that he had retreated to jokes of that caliber was as good as waving a white flag. "I really am serious, Alan. I'm sorry to be brutal, but I really don't have the time to be subtle. I intend to do this, and I want you to be good about it—not just now, but afterwards and forever—because it's no mere matter of making me feel as well as can be contrived while my brain still works. A Human Tree is forever; I'll be part of your life, and Jeanie's and Mike's lives, until the day you and they die. Even if you were to dig your heels in and refuse to have anything to do with me, I'll still be there."

"No you won't, Sarah," he said, his infernal stubbornness drawing a reluctant Parthian shot out of his determination to surrender and be kind. "They might be right, technically, about the continuity of cellular life and the preservation of the fundamental DNA-complex, but it won't be *you*, any more than a corpse in a grave or an urnful of ashes in the trophy cabinet would be you. You'll be dead, Sarah—if you don't fight."

"I'll be dead whether I fight or not," Sarah told him, gasping again as cymbals suddenly joined in with the climax of a crescendo of screeching violins and booming brass. "You know that. I *have* fought—but I didn't win. That Age of Miracles never arrived."

He said nothing, but she went on, for her own sake, externalizing the proud of her imagination by way of paying back the debt she owed the trees and the road. "The human body simply isn't built in such a way that we can stay in the prime of life, incapable of permanent violation by disease or injury, for centuries. Transfiguration is the only possible immortality. It's not enough—nobody ever claimed that it is—but it's what we've got, and all the evidence suggests that it's the best we'll have for quite some time."

The prosecution still had nothing to say.

"Maybe you're right," Sarah continued, "and there's no essential difference between being planted and being buried or cremated—but even if that's so, the choice between the three is still a meaningful one. You have to respect my choice, Alan; you have to help Jeanie and Mike make the most of it. You have to preserve the meaning of what I'm doing, even if you do insist on thinking, in your heart of hearts, that I'm dead and gone and that the Tree is just an insult to my memory."

She knew that her voice had expressed her pain, in spite of the road and her best intentions. Alan put his hands up to his face, and covered his eyes with his fingers. It wasn't the long straight tunnel through the Neogymnosperms that he was refusing to see, and it wasn't the automatic pilot's ultra-careful driving of which he was despairing.

After a couple of minutes he put the hands back on the wheel again, not because he wanted to pretend to steer but simply because he needed to get a grip on *something*. "I wish the highway weren't so goddam *boring*," he said. "I wish Transfigured trees weren't so goddam *orderly*."

"We'll be back in the Halo in no time," she told him. "Urban design is the etiquette of New Global Civilization. The days when cities and their suburbs just *sprawled* are gone forever."

"You've never been to Denver or Chicago, let alone New York," he retorted, for no particular reason. "We're lucky, living way out west. Arizona's still on the real frontier, you know, even if it's been Transfigured out of all recognition. It's one of those places where people get right on and *do* things. No inertia, lots of *fight*."

"And where are the biggest Human Forests in the USA?" Sarah was quick to say. "California, inevitably. Oregon, of course. Montana and Wyoming are the third and fourth. All pioneer country: join up the dots and there's the frontier of the future."

"It'll be a hell of a long time before anything bigger than a Human Copse sprouts up in Utah," Alan opined.

"No, it won't," Sarah insisted, as gently as she could. "Sacred Groves will be everywhere before you know it, linking up from sea to shining sea. It really will be a New World, Alan—and I'll be part of it. I really will."

* * * * * * *

Jeanie and Mike were well past denial and anger by now. Unlike Alan, they'd already made their psychological adjustments to the verdict and sentence that their parents brought home. They backed Alan up when he said that Sarah really ought to go to bed, but she stood firm—or, to be strictly accurate, sat firm—on the living room sofa. The sofa was directly opposite the painting over the mantelpiece that displayed Old Arizona in all its ancient glory, all desert and bare rock, glowing sulky red in the setting sun. In Old Arizona, if its current iconography could be believed, the sun had always been setting.

The children weren't in the least surprised or dismayed when Sarah told them that a Forester would be calling round to make arrangements; Jeanie was ten and Mike was seven, so they'd both grown up with Human Forests as a fact of life, and the fundamental notion didn't seem in the least strange or alien to them. That didn't mean, though, that everything went smoothly.

"You have to be planted in the yard," Jeanie said. "This is our home. You have to stay here, with us."

"That's not as good idea, my love," Sarah told her. "Maybe if we were feudal barons living in English castles we'd be able to think of our homes as long-standing family heirlooms, but we're not tied down, and we shouldn't want to be. This is the only home you've known, so far, but it certainly won't be the only one you'll ever know." She had to stop then, but she fixed Alan with a commanding stare.

"That's true," Alan admitted. "Ours is a land of opportunity, and you have to be free to take those opportunities when they arise. When you go to college, and get jobs, you need to be free to go where you want and need to be. Having your mother's Tree in the back yard is an anchor you can do without."

He stopped there, leaving Sarah to add: "Besides which, Trees belong in Forests. They thrive in company and don't do so well in isolation. It's best if I'm somewhere where I can belong, where you can visit me when you want to, without ever getting to take me for granted."

"You'll still be able to talk to us when you're a tree, won't you?" Mike said. "You'll still be able to listen to us."

"No, Mike, I won't," Sarah told him, taking him by the hand. "I know you've seen a lot of cartoons that represent Animal Trees as things with eyes and mouths, which wave their branches around as if they were arms, but it's not like that. That's just a joke. When I start the series of injections—by which time I'll be in the hospice—I'll go to sleep, and that'll be the last time you see my eyes or hear my voice. Transfiguration takes a long time. It'll be months before I'm ready for planting, and more than a year before I begin to look much like a tree, but once I do, I'll be a real tree. I won't have a brain any more, so I won't be able to think, let alone talk or listen."

"But you'll be able to dream," Jeanie put in.

"No, I won't," Sarah said, determined to tell the truth, the whole truth and nothing but the truth, as Alan would expect and was entitled to demand. "That's...well, not a joke, more a myth. People like to imagine Human Trees being in a kind of dream state, but once my brain's gone...."

"Where will it go?" Mike wanted to know. He was frowning; perhaps he had been taking it for granted that the human body was still contained within the Human Tree, like a mollusk within a shell.

"It will change, just like everything else," Sarah told him. "It will change into the flesh of the tree."

"But if you can't dream," Jeanie said, hesitantly, "how can you be in vegetable heaven?" Sarah knew that Jeanie didn't really believe in vegetable heaven, but the dutiful ten-year-old probably thought that she ought to make an effort to conceive of it in terms of some kind of hallucinatory state, just in case her mother expected to arrive there.

"If there is such a thing as vegetable heaven," Sarah told her, choosing her words carefully, "it's not something you have to dream. It's just a matter of *being*."

"How long will it take you to pay your carbon debt?" Mike asked. Even though the Ecocatastrophe was over, elementary school children were still taught about the carbon economics of everyday life as a matter of routine. Mike couldn't put numbers into the equations—he didn't know how much carbon dioxide a flight from Phoenix to Miami or a transatlantic trip by a container ship would pump out—but it had been drummed into him that his remoter ancestors had been villains because of the awful extent of their carbon footprints, while his more recent ones were heroes, by virtue of their tip-toed ingenuity, thus bringing the historical account-books into a belated but triumphant balance.

"It's not as simple as that," Sarah told her son. She wanted to explain him that it wasn't sensible to divide carbon debt up into individual slices, because it was the economic activities of the whole society that produced the greenhouse gas surplus, and that distributing blame between different nations in the fashion that had made the USA the Great Green Satan of the 21st century was patently false accounting, because the economic activity of a particular nation had to be seen in the context of the global economy, but she couldn't have done it, even if there had been any point. There was too much Balakirev going on. What she actually said was: "It isn't a matter of one Tree paying off one person's debt; it's a matter of whole Human Forests making their contribution to the work that's done by all the other Transfigured Forests in the world."

Alan snorted, but when Sarah stared at him he pretended that he had only been stifling a sneeze.

"One day," Jeanie said, abandoning questions in the interests of demonstrating her intellectual superiority over her younger brother, "there won't be any other Transfigured Forests. All the world's forests will be Human Forests, and then we'll really be Responsible. *Then* we'll have paid our carbon debt to the Earth."

Sarah was glad that Alan hadn't saved up his snort, because Jeanie was old enough not to be fooled by any kind of belated bluff and would have taken it personally. She resisted the temptation not to correct her daughter's Utopian excess, even though she could see that her red-faced husband was biting his lip as he forced himself to maintain diplomatic silence.

In her heart of hearts, Sarah hoped that Jeanie might be right, and that there *would* come a day—if not for thousands of years—

when *all* the Transfigured Forests had been replaced by Human ones of every race and nationality, so that all the human beings alive throughout the world could live in the perpetual company of their ancestors, and the species really could consider itself Reverent and Responsible. It was not a dream that she would be able to maintain when she became a Tree, but it seemed to her a legitimate hope that she might live to see such an era, albeit not in her present frail and feeble form.

"What *kind* of tree are you going to be?" Mike asked, his mind on more down-to-earth matters. "I don't want you to be a Joshua tree or a monkey-puzzle."

"Mummy doesn't get to *choose*," Jeanie put in, getting slightly carried away with her own supposed expertise. "Human Trees are Human Trees, not any other sort. They're evergreen, but not like Lollipop Pines. They're just...themselves."

Sarah didn't want to complicate the issue by arguing that the present uniformity of Human Trees was probably just a phase, and that all kinds of choices might have opened up by the time Jeanie had to decide between planting and death. In any case, there was no choice at all for *her*. Her fate was sealed, to the sound-track of *Scheherazade*.

* * * * * * *

The Forester's name was Jake. He was a little too smartly dressed, as if he were overcompensating for the image most people had of Foresters, but he wore his blond hair long and curly, obviously thinking of it as a precious asset not to be sacrificed on the altar of businesslike appearance. He arrived on time, but Sarah had made a late appointment so as to be sure that Jeanie and Mike would be in bed, so it was after dark. She still wasn't absolutely sure that Alan would be able to contain himself in confrontation with Personified Temptation.

To start with, though, Alan was on his best behavior. He meekly poured out a glassful of apple-juice when Jake explained that he didn't drink alcohol or coffee, and gave him the benefit of the better armchair.

"I've brought you the standard literature," Jake said, handing Sarah a sheaf of leaflets with which she was already perfectly familiar, "but I'd like to give you a brief verbal explanation of what will happen. It helps clarify matters, in my experience, and brings questions to the surface that each particular individual needs to ask."

Sarah could read her husband's mind well enough to know that the words *pompous* and *asshole* must be drifting through it, but Alan said nothing, and Sarah consented herself with an encouraging nod.

Jake launched into his spiel. "The Association of Human Foresters," he said, "is not a commercial organization. No one ever set out to exploit this particular corollary of Transfiguration technology for financial gain. No one ever tried to sell it to the public by advertising. The AHF was summoned into being by public demand.

"The scientists who developed the metamorphic techniques that allowed living plants to be Transfigured realized almost immediately that animals might be Transfigured too—only into plants, of course, not into other animals—but they thought of it as a technical challenge rather than a practical enterprise. The earliest animal Transfigurations were all done in a spirit of pure experiment; it was a big surprise that it worked so well, after all the disappointments of the past in respect of animal genetic engineering.

"Once it became popular for people to preserve their pets by Transfiguration, the demand for Human Transfiguration became increasingly insistent and urgent. The legislation went through with extraordinary ease. The AHF was created and regulated by state governments working in association with HMOs and a number of existing charitable organizations. I won't bore you with the bureaucratic details; suffice it to say that I'm not here as a salesperson or a social worker, but merely as someone answering a summons. Our services do have to be purchased, but a part of the care we give is reclaimable through health insurance, and the remainder compares favorably with the average costs of interment or cremation.

"What will happen, Mrs. Whitney, if you decide to go ahead with the Transfiguration, is that when you and your supervising physician decided that the time is ripe—which shouldn't be problematic in your case, given that the progress of your cancer can be accurately monitored—you'll go into the hospice. The first injection is administered on the day following admission. You'll slip into a coma almost immediately, and the Transfiguration will begin. The first injection contains the first of three suites of viral organelles that carry the extra genes you'll require in order to complement your own DNA, but its main components are the catalyst for the despecialization of your own cells and the trigger-proteins for the construction of the tuberochrysalis. The tuberochrysalis takes seven days to form; it will retain the basic outline of your bodily form, but many of the individual features will be lost. The second suite of supplementary DNA is injected on day eight, the third on day twelve, and the final batch of catalysts on day sixteen—by which time all your reblastularized cells should have been infected by the vectors. There's a long lag phase thereafter, but planting out of the tuberochrysalis is usually practicable after forty-five to fifty days, if no complications set in."

Alan's patience finally snapped. "What complications?" he asked, sharply. "Your brochures don't say anything about goddam complications."

"There's a good deal of detailed data available on-line, Mr. Whitney," Jake replied imperturbably. "The complications so far observed are various, but uncommon. The worst case scenarios involve the rejection or suppression of one or more of the supplementary DNA packages, which can prevent the completion of Transfiguration if the situation isn't remediable. Usually, it is. To date, more than ninety per cent of our Transfigurations have been completed without any complications at all, and more than ninety-nine per cent have been brought to a satisfactory conclusion. Given the improvements we've made in our techniques and our understanding, we expect those figures to rise in the future, steadily if not sharply."

"I've studied the probabilities," Sarah said, more to Alan than Jake, "I understand the risks. There's something less than a one per cent chance I'll end up dead—as opposed to a hundred per cent certainty that I'll end up dead if I don't opt for Transfiguration."

"The figure is a hundred per cent either way," Alan retorted. "The difference is that if you opt for this travesty they kill you themselves instead of your dying a natural death, and then they turn your remains into protoplasmic mush before getting their ninety-nine-percent chance to turn that mush into a tree."

Jake opened his mouth, but Sarah held up her left index finger to silence him from a distance. "You're a servant of the law, Alan," she said. "You're no longer allowed, let alone obliged, to disapprove of assisted suicide in cases of terminal illness. Choice and painless injection beat natural death hands down, and there's nothing scary or obscene about protoplasmic mush. What do you think decay does to a body? What do you think happens to its carbon atoms thereafter? Whether they travel via the guts of graveworms or dissipate into the air as crematorium smoke, they eventually wind up as the transitory flesh of plants. All the Foresters are doing is bringing some order and focus to the process, cutting out the middlemen and maintaining the continuity of cellular life."

"The *illusion* of the continuity of life," Alan objected.

"It's not an illusion, Mr. Whitney," Jake said, quietly. "Nobody pretends that it's any kind of continuity of consciousness, but it *is* a continuity of life, literally and materially. It's not for me to say whether soul and spirit are illusory, or where they reside if they aren't in the flesh—but there really is no doubt that a Human Tree really is an extrapolation of human life. Your wife's Tree will include the full complements of her nuclear and mitochondrial genome, biochemically organized in exactly the same way. It's just that some new genes will be added and different genes expressed, in different combinations, in different kinds of specialized cells."

"That's three dimensions of difference," Alan pointed out, "and that's all the difference in the world. This is just another scam, in the great tradition of cryonics, offering the shadow of a perverted hope where none really exists."

"No, Alan, it's not," Sarah told him, pursing her lips and moving to the edge of the sofa because her intestines were being attacked by a gang of frenzied cellists. "There's no promise of resurrection here, false or otherwise."

"Becoming a Human Tree is a matter of moving on to a new phase," Jake added, trying to catch the enemy in a pincer movement. "Of course there's no way back, any more than there is from death—but the individual remains instead of being broken down into component molecules and redistributed as raw materials."

"*But that doesn't mean that it won't be me*, Alan. It'll just be a *different sort of me*, here on Earth and not in any kind of heaven. That's where I want to be, Alan, That's where I want to *go*."

Jake had another leaflet about that, offering a choice of Human Forests, woods and private plots. Sarah had already made up her mind. She wanted to be up on the Colorado Plateau, partly because the Grand Canyon Surround was further on the way to becoming a mature forest than most, and partly because there was more rain up on the plateau than there was in the reclaimed Mojave. The plateau was a little further away, but it seemed to her to be the right sort of place for the kind of creature she was ambitious to become. The Halo of Phoenix Reborn was a fine place for humans to live, but Human Trees had different needs, and ought to have different delights. Even more than the Sierra Nevada, in Sarah's opinion, the part of Colorado plateau that lay south of the Utah border foreshadowed the future of America—not the future of destiny, which really was a silly illusion, but the future to which the Foresters were attempting to act as midwives: the future in which civilized humans would live in the carefully-maintained interstices of an Eden of Human Trees that were both Trees of Knowledge and Trees of the Knowledge of Good and Evil.

Alan's mind was on other things. "What about me, Sarah?" he complained. "What becomes of me, while you're turning yourself into some vast lumpen tuber, and growing into something alien?"

"You have two children to bring up, Alan," she said, more sharply than she intended, because she was hurting like Khatchaturyan, but less romantically. "You have to hold yourself, and them, together. When that job's done, you're free. If you don't want to be planted, that's up to you. You can send your carbon atoms into the future along any trajectory you choose. The only thing you can't do is abstract them from the great chain of being. One way or another, they'll live again."

"*They* don't live at all," Alan insisted. "*Cogito, ergo sum*; real life is made of thought."

"Don't be silly, Alan," Sarah said. "There are countless plant and animal species that don't think, but they're all *real life*."

"A lot of husbands and wives," Jake put in, "are planning to be planted side by side. For the present, at least, Human Trees are monoecious; male and female flowers are borne on separate dendrites. Thus far, they're all sterile, and their capacity for vegetative reproduction is minimal, but the science and technology aren't standing still." He stopped, apparently realizing that both his clients were staring at him, no longer divided.

"Are you suggesting that we ought to be thinking about tree sex and tree children?" Alan asked, stonily.

"Some people do," Jake said. "Some people seem to get quite a thrill out of the idea—but no, I'm not. Fruitful exchanges of pollen might some day be possible, but Mrs. Whitney's Tree will be unable to produce fertile seed. What I *do* suggest you might think about, though, is togetherness. Skeptics insist that love can't transcend Transfiguration any more than it can transcend death, but there's a sense in which it can. What I'm saying is that intimate contact can be reconfirmed and reconstituted within a Forest. Nor will Human Trees remain the end-point of our potential existential journey for very long. As I said, the science and the technology aren't standing still."

Alan was silent for a few seconds before he looked at his wife and said: "I don't know how far I can follow you down this road, Sarah. It's your choice, and I have to go along with it, for the kids' sake—but I can't make you any promise about joining you when my time comes."

"I never expected that you would," Sarah said. "After all, you'll still have a life to lead when I'm a Tree, and there'll be other people in it."

"In a Forest..." Jake the Forester began—but this time, both his listeners raised their left hands, with the index fingers extended in exactly the same fashion.

A further half-minute passed while Sarah and Alan looked at one another. Then Sarah turned away, satisfied that everything that could be settled had been settled. "I'd like to get the paperwork done now," she said to Jake, "and set the wheels in motion." Mercifully,

Mussorgsky was now experimenting with one of his quieter movements.

"Certainly," said the man who wasn't a salesperson, his blond hair rippling as he nodded his head.

****

When Sarah could no longer get out of bed her doctor increased the dosage of her pain-killers to the point where the Russian symphony was transfigured into a French piano concerto, all the way to mere Debussy. She was still able to postpone her dosage if she waned to retain a greater clarity of mind or a while, to make her final reckonings with her parents and former colleagues, but the Russians had gone for good; her most attentive times, during the last weeks of her life, were mostly Bach and Brahms, with only occasional interventions of the Blues.

Three months passed before she moved from her home into the hospice, and began the next phase of her existential journey.

Sarah was tempted, then, to come off the diamorphine altogether, in order to experience the preliminary stages of her Transfiguration as fully as possible, but it couldn't be done. The withdrawal symptoms would have wrecked her perceptions far more comprehensively than the drug itself. She had no alternative but to drift off in a haze, the authentic music of her life fading into a background of gentle crooning and soft swing.

She lost track of time before she would have wanted to, if she had still been able to want anything in the final weeks. Grief, sorrow, regret and the other members of their dysfunctional family were long gone, and didn't even pop in to say goodbye while she was in the hospice.

She tried to talk to Alan, Jeanie and Mike while her vocal cords still worked, but her conversation had evaporated, leaving nothing behind but a crusty residue of platitudes that didn't sound like her at all, but seemed to belong to some vapid ghost strayed from the electronic hinterlands of daytime 3-V. There would have been nothing left of her old self at all, even before she moved into the hospice, if it hadn't been for the dreams. The diamorphine couldn't keep her properly awake, but it did lend color to her sleep.

In a kinder or fairer world Sarah might have been able to dream the kind of dream that Jeanie had sketched out for her: the dream of a future Golden Age in which the people of every nation on Earth would cultivate their own customized Edens, where their haloed hi-tech cities would be ringed and separated by ancestral forests, and the soil of the United States of America would be bound together by the roots of its people—not its makers, let alone its original natives, but the roots of its *remakers*: the movers and the shakers who had finally brought it to its proper constitution and its true destiny.

In fact, the dying Sarah never had that kind of dream. She didn't dream about history and destiny at all. Her imagination had retreated to a smaller scale. She dreamed about her children, a little, and her husband, a little more, and other people a little more than that, but mostly she dreamed about numbers and balancing accounts.

The dreams weren't nightmarish, as they might have been if the figures had refused to add up and the failure of her enterprise had generated panic, but that hardly ever happened. Almost invariably, the figures did add up, so invariably that there was no sense of triumph in their settlement, but not so easily as to rob her of all legitimate satisfaction.

Then she yielded to the exotic pressures of the injections and was Transfigured into a Tree, in which form she lived for a further thousand years before embarking upon the next phase of her technologically-assisted existential journey.

As to whether Alan, Jeanie and Mike came to visit her often, or whether they interrogated her, or whether any of them eventually joined her in the Forest, Sarah's Tree had no idea. There was no way she could know and no reason why she should; it was their business, and they were free to make what use of her they could, or not.

Sarah's Tree was no longer capable of dreaming, but all the sensitive creatures that were able to hear her, in the course of her millennial existence, perceived that the song of the New World's wind in her leaves and branches was infinitely more beautiful than silence.

SLEEPWALKER

"Never volunteer", they say in the army—well, they say it in the poor bloody infantry, if not in the officer's mess. It's good advice, in its way. What it means is *take no risks*; be satisfied with what you have, be it ever so humble. The problem is that progress requires risk-takers; it depends on the willingness of unreasonable men to be dissatisfied with what they have, and to volunteer as pioneers of the great unknown.

I've always been an unreasonable man. "He'd rather be wrong than orthodox" is what they used to say about me. I don't know what they say now.

It didn't seem like such a big risk at the time. I knew all about Jouvet's research, of course, and I'd always been intrigued by it. Surgical removal of a body called the pons from a cat's brain takes out the censor which switches off the motor nerves while the animal dreams. Pons-deprived cats act out their dreams; their sleep-life becomes manifest. It was obvious, of course, that the same effect could be obtained without actual surgery, if only one could learn the trick of it.

People who talk in their sleep are acting out one aspect of their dreams, after a modest fashion. Sleepwalkers are acting out more aspects of their dreams, in a slightly less modest fashion. So it wasn't that much of a shock when Spicer came to me and said: "We've figured out how to do it. Temporary chemical interruption of the censor in the pons. We can get people to act out their dreams in full—all we need is volunteers." Which, roughly translated, meant: "How about it, sucker?"

I said *yes*. What's so terrible about the thought that you might act out your dreams while being closely observed by a battalion of psychologists? After all, even if I dreamed that I was doing something reprehensible, like committing murder, I wouldn't actually be doing it. Jouvet's cats dreamed of catching mice, but the mice weren't actually there for them to catch—they were imaginary mice, entirely in the eye of the dreaming beholder. It did occur to me that there were things one does in dreams that might be slightly more embarrassing than committing imaginary murder, but in the cause of science one has to be prepared to risk suffering a little embarrassment now and again.

As things turned out, that wasn't the problem. At least, it wasn't the *whole* problem.

The most interesting thing that Jouvet's research revealed, of course, is that feline dreams are so damnably *sensible*. A cat's dreams provide an arena in which instinctive behaviors can be practiced and commonplace mental routines enhanced. Human dreams aren't like that. Human dreams are much more bizarre and much sillier. One popular theory holds that that's the case because humans don't have very much in the way of inbuilt instinctive behavior; in consequence, the human dream arena is redundant inner space that has run to dereliction. Humans don't need to rehearse inherited patterns of behavior, so they just have this empty stage where all kinds of rubbish drifts around, accumulating in untidy heaps. Perhaps it's true; I don't know whether my own experience favors the hypothesis or not. I only know that it doesn't really matter.

Spicer's drug worked. It really did cut out the censor in the pons on a temporary basis, with no harm done—no physiological harm, at any rate. Unfortunately, the preliminary experiments with cats and rats didn't show up one interesting side-effect that was only applicable to humans.

Because of the way in which cats use their dreams, they need to remember them—mental rehearsal is no use if it's all forgotten. Humans not only don't need to remember their strange and silly dreams; it would be a positive disadvantage if they did. Humans, in consequence, have a double censor built into the cytoarchitecture of the pons, which not only inhibits motor activity but memory as well.

Spicer's drug switched off the whole thing. Not only were he and his team able to watch me acting out my dreams; I was able to remember them, in every detail, exactly as if they had been lived experiences.

When Spicer and the team first realized this, of course, they were overjoyed. After all, there's only so much you can learn about a dream by watching it being acted out. They only had half of every dialogue and they couldn't see the other entities to which I was reacting. To them, the memory retention seemed like an unexpected bonus—and it was. It was a bonus for me too, but not in exactly the same way. Perhaps it's more than a bonus. Perhaps—potentially, at least—it's a great boon to humankind, or at least to that fraction of humankind which has the capacity to cherish its dreams and learn from its nightmares.

I used to have a life. One lousy, linear life. One incredibly straightforward, utterly ordinary, *everyday* life. Not any more. Now I have a hundred lives, and a thousand more to look forward to.

I used to be a citizen of the world, but now I'm a citizen of the multiverse.

I used to be a glorified lab rat with only half a brain, but now I'm a king of infinite space, and I'm using my brain to the full.

I still live in what you ponsed-up paupers call "real life", but I also dream, and I also live in my dreams. Like everyone else, I'm the product of my memories, but there's a great deal more productivity in my current make-up than there used to be, and things can only get better as I become more expert in the art and craft of dreaming.

Dreaming *is* an art. And a craft. My journey has a long way still to run.

Of course I have bad dreams occasionally—who doesn't?—but even the worst of them can be savored, knitted into life's rich tapestry.

These days, I can hardly wait to go to sleep, and the biggest bummer of every day is waking up. It's waking up that brings me down, reducing me to what I, just like everyone else, used to think of, in the pathetically narrow fashion dictated by the tyrant pons, as "real life".

That pathetically tiny fraction of my existence seems nightmarish nowadays, because it's so dull and predictable and so utterly banal, like a mental rehearsal for death.

Since my dreams became *real experience*, as tangible and meaningful as any other, I've become ten or a hundred times the man I used to be.

And that's why I can't understand why Spicer and the rest of you want to take me off the stuff.

So what if I have been stealing from the store? So what if I have been sneaking off to take naps at every opportunity? Can't you see that I'm in pursuit of a better and fuller life, and that what you're perversely intent on dragging me back to is sheer hell? Can't you see that I'd do *anything* to preserve what I have now?

If you want my advice you can have it. "Don't knock it until you've tried it" is what I say. Volunteer. Learn to sleepwalk. Become the master of your own fate.

Do you want to be in the poor bloody infantry all your life?

THE BEAUTY CONTEST

The school counselor was running late, so Janice turned up for her eleven-thirty appointment to find Helena still waiting in the corridor for her eleven-fifteen.

Ordinarily, it wouldn't have been a problem, because she and Helena had been friends ever since year four, when they'd still been at primary, but since they had both fallen in love with Crispin Slipman they had virtually stopped speaking, so it was a little awkward. Janice felt that she'd somehow got drafted into a beauty contest that she was sure to lose, given that Helena was blossoming so conspicuously. She knew that the chances of a hormone factory like Crispin learning to look beyond mere appearances to search for the inner beauty behind the mask were pretty slim.

The silence lasted seven minutes before it became unbearable.

"Who's in there?" Janice asked, figuring that it was better to behave in a civilized manner than continue to suffer the deep embarrassment of mutual antipathy.

"Timmy the geek," Helena replied, tersely.

"That must be why she's running late," Janice said. "He's probably explaining his profile to Mrs. Charters—which is bound to take five times as long as her explaining it to him."

Helena hesitated for a moment, but apparently decided, somewhat to Janice's relief, that it would be safe to converse as long as the topic was harmless. "Given that the whole of year nine has to go through this farce in the next two days, you'd think he'd just say that he didn't need any advice and get the hell out of there," Helena said, wearily, "but not our Tim. We should be okay, though—it's the next two in line who'll be in danger of missing lunch."

"They'll get rescheduled," Janice predicted, confidently. "Mrs. Charters never misses lunch."

That brought a weak smile to Helena's face, although Mrs. Charters wasn't as conspicuously overweight as poor Mrs. Bunn—but it wasn't enough to defrost their chilled relationship.

"I'm not going to be in there long," Helena stated, forthrightly. "I don't want to listen to a long list of the horrid diseases I might get and the precautionary measures I ought to take. It's one thing to wear a chip so the paramedics know what to do if I ever get knocked down, but it does more harm than good to make us all paranoid about things that haven't happened yet and might never happen at all."

"There might be good news as well," Janice said. "Libby Leigh from year eleven got a contract with a pharmaceutical company because she's got some rare gene that makes her eggs worth a fortune. She'll be off to private school next term."

"That's even worse," Helena said, flatly. "I'm not letting anyone cut me open to get at my eggs, no matter what sort of treasure turns out to be buried in my genome."

It wasn't the kind of prospect Janice could relish, either, but she knew that Mum and Dad were really struggling with their repayments now that Mum was on short hours, so that any kind of windfall would be a godsend, even if it did involve the excavation of her ovaries. She knew how unlikely it was that her DNA profile might be commercially exploitable, though—the odds weren't much better than landing the jackpot in the lottery. Although the TV ads harped on about the value of self-knowledge and lifestyle adaptation, everybody knew by now that what most people got out of the National Genetic Database was a set of dire warnings about hazards of which earlier generations had remained blissfully unaware, and a very long list of things they'd be better off not doing, no matter how much fun they'd got out of them until now. If you were lucky, the only strong advice you got was to avoid certain foodstuffs, but Janice found it difficult enough to stick to a supposedly sensible diet without having any more restrictions piled on top.

The door to the counselor's office opened, and Tim Sillington came out, shaking his head. He started slightly when he saw Helena and Janice waiting, and remembered just in time to hold the door open so that Helena could go through. When he closed it behind her he was alone with Janice—a prospect which he evidently relished more than she did. There wasn't anything particularly unattractive about him in the physical sense, although he was too short, his hair was always in a mess and he looked as if his mother shopped for him exclusively in Oxfam, but he had been "Timmy the geek" ever since he first set foot in the school. Janice wasn't the sort of girl to welcome attention from a designated outsider. Tim's eagerness to help Janice in Computer Science classes had become a matter of general comment, far more embarrassing to her than her obvious need for assistance.

"This is a complete waste of time," Tim told her.

"I thought you'd be all in favor of it," Janice said, surprised to hear such a judgment from his lips. "You're the last person I expected to argue that the NGD won't change anything."

"I didn't mean the NGD. *Of course* it will change things—it'll change *everything*. I'm talking about the Stone Age software Mrs. Charters is using to analyze the results and churn out advice. Listening to her, you'd think it was all about vulnerability to diseases and possible adverse drug reactions."

"Isn't it?" Janice asked, before she could stop herself.

"Of course not. I'm not denying the usefulness of pharmacogenomics, although the hype about personalized medicine is way over the top, in my opinion, but there's so much more in there, if people would make the effort to cover all the angles. You'd think the school would be sufficiently interested in bringing out people's latent abilities to invest in software to analyze facilitating genes, but apparently not. I can see why they're a bit wary about elementary correlation programs, but they're just a convenient starting-point for more serious investigation. If schools aren't in the business of trying to identify and foster latent talents, who is?"

"We did those career aptitude questionnaires last year," Janice pointed out.

"Exactly," Tim said, contemptuously. "The NGD's finally up and running, and the school is still trying to figure out what people might be good at by asking stupid questions about what kind of pet they prefer and what magazines they read, which no one with an atom of sense answers honestly. They might as well use astrology. What sign are you?"

"Capricorn," Janice replied, automatically.

"There you are. I bet everybody in the school knows their birth-sign. Three out of four probably read their horoscope ever week, and two out of three probably know what sign their ideal partner is supposed to be—but half of them claim not even to want to know what the real implications of their actual innate composition might be. Taking notice of questionnaire-based personality profiles was always stupid, but now that we have a chance to analyze *real* potential it's so perverse as to be insane. I even volunteered to provide the school with copies of the most recent genome-analysis software, and all old Charters could say was that she'd pretend she hadn't heard me."

"Why?" asked Janice, quite mystified.

"Because they're bootlegs, of course. She thought I'd stolen them from some pharma company's secret files. I wish! I can't even hack into the police computer, let alone a major biotech company—and it's not for lack of trying. I got them from a friend of a friend."

When Janice made no reply to that, Tim paused for breath, and then said: "If you want to know the whole story, you could bring your chip round to my house. I'll run it through everything I've got."

The offer caught Janice off guard, but she managed to hesitate long enough to formulate a considered reply. Eventually, she said: "That's okay, thanks. I probably won't be able to understand what Mrs. Charters has to say, let alone anything complicated."

"I can interpret it for you," he was quick to say.

"No, honestly," she said. "I really haven't time." Partly to deflect his attention away from her, and partly in response to a mischievous impulse, she added: "You might ask Helena, though. She was just complaining about the whole thing making her paranoid. She might appreciate some good news to balance out the stuff about which diseases she's most at risk from and which foods she'd do better to avoid."

It was Tim's turn to hesitate. "Oh," he said, "Well, maybe. I'll have to catch her later, though. I'm running late."

Aren't we all, Janice thought, as he turned on his heel and hastened away. She hadn't known before that Tim was afraid of Helena, although now she came to think about it, Helena did tend to tease him quite a lot. Then again, Helena also teased her, especially about needing help in Computer Science, and Janice tried not let it bother her, even though it had become more difficult since Helena's more obvious attractiveness had begun to exert its magnetism on Crispin.

Once she was alone again, though, Janice began to think about what Tim had said, about "facilitating genes" and "correlation programs". She had only the vaguest notion of what the phrases might mean, but she had heard enough of Dad's rants about the evils of "genetic fortune-telling" to know that there were people selling software that was supposed to discover people's "hidden talents" by analyzing their DNA profile chips, and that lots of dating agencies were advertising their ability to make matches on the basis of "genomic compatibility". Dad thought they were all frauds and conmen—just like the astrologers he condemned with a contempt even greater than Tim's—but Janice wasn't so sure. After all, if the National Genetic Database was going to make it so very difficult for anyone to commit undetectable crimes, it certainly ought to be able to help people figure out what kind of jobs they'd be best suited for.

She was sufficiently taken with this idea to ask Mrs. Charters about it when she finally got her turn in the comfy chair—by which time there were two more year nines queuing behind her, fretting about the awful danger of not getting to the canteen while the chips were still hot and crisp. Helena's promise to be brief had not been honored, although the black expression she had been wearing as she stomped off suggested that the delay had been the counselor's fault.

"You've been talking to Timothy," Mrs. Charters said, accusingly, in response to Janice's question. "That boy's headed for trouble."

"He doesn't really try to hack into the police computer," Janice said, defensively. "He just says that."

"You don't have to make electronic transfers from other people's bank accounts to get into trouble with the law," Mrs. Charters said. "All copyright violation is theft. But that's not the point. The point is that the school only uses *reliable* software packages to analyze its pupils' genomic profiles, and only dispenses advice of proven medical significance. It's one thing for me to advise you on a suitable diet and make sure that you understand the contraindications of common prescription drugs, quite another for me starting speculating about *loose correlations*."

"But you do career advice too," Janice pointed out. "You even talk to all the year elevens about which subjects they might want to specialize in at A level."

"On the basis of their declared interests and demonstrated aptitudes," Mrs. Charters said. "Not on the basis of how many gene-variants they happen to have in common with the average professor of physics or the average busker in the underground. That sort of thing is no more scientific than systems for betting on a roulette wheel based on the sequence of numbers it's already generated—it's a matter of searching for haphazard coincidental patterns of no real causal significance."

Janice daren't ask what that meant, in case she was delayed as long as Helena. She just nodded her head and looked expectant.

"Well," said Mrs. Charters, "I'm glad to say that your profile appears to be as close to problem-free as any mere human is likely to get. You're carrying seven recessive genes that would be lethal if they weren't coupled with healthy dominants, but none of them is likely to pose any medical problem, even in cells where the dominant is disabled by mutation, because the defective cells will be purged rather than hanging round to cause trouble. You have six genes associated with degenerative diseases, but none that's likely to take effect until you're at least sixty years old—by which time all six conditions will probably be treatable. Your metabolic system appears to be in good shape, so there are no ordinary foodstuffs likely to cause problems, and you're highly unlikely ever to require any currently-prescribable medical treatment to which you'd have an innate adverse reaction—although you'll have to bear in mind that the data in your profile can't take account of environmentally-induced allergies. Do you understand what I'm saying?"

Janice knew better than to say no, but she knew that if she simply said "yes" Mrs. Charters would ask her a probing question, so she had to contrive a more intelligent response. "You're telling me that I'm lucky," she said. "You don't have to tell me to stop doing anything, or to be wary of anything that I'm likely to run into."

"That's the gist of it," the counselor admitted, with a slight sigh. "You're one of the favored few who won't actually need to understand any more than that. Run along, and sent the next one in."

* * * * * * *

Janice knew that she ought to have been doubly pleased, not only because her DNA profile had exposed no major problems, but because that meant she didn't have to grapple with the problem of trying to understand any problems it had exposed. All she could think of, though, was the length of time Helena had spent in the counselor's office and the desperation on her face when she'd come out. It was obvious now that Helena must have been dismayed by something worse than mere delay—must, in fact, have been doubly unlucky. Not only must there have been something worrying in her profile, but Mrs. Charters must have added insult to injury by making every effort to ensure that Helena understood exactly how worrying it was.

When Janice got back to Mr. Thompson's History class she saw that Helena was still looking grim, but she had no chance to commiserate. The bell went almost as soon as she got there, and Mr. Thompson called her to his desk so that he could fill her in on what she'd missed. He hadn't been doing that for everybody, of course, but what Janice had missed had included the homework assignment, and he certainly didn't want her to have any excuse for not doing it.

By the time Janice got to the dining-room Helena was already deep in conversation with Crispin, and the other seats at their table were all full. In fact, almost all of the seats at most of the year nine tables were full; by the time Janice had filled up her tray and handed in her ticket it was a straight choice between sitting opposite Tim

Sillington or joining a company of year sevens. The latter being unthinkable, she took her place at the end of the year nine dregs' table, trying to be grateful for the fact that she was able to leave an empty seat next to her because Mrs. Charters still hadn't released Clive Pinks.

"How did it go?" Tim asked her.

"It's supposed to be confidential," Janice pointed out, before relenting and saying: "Fine. No buried treasure, but no problems either."

"That's good," he said. "Except that Mrs. Charters' programs only deal pick up the most obvious treasures—the gold and silver, but not the diamonds or rubies, as it were."

Janice looked up at him then. "You mean that I might have something saleable in my genes that the people compiling the National Database missed?"

Tim blushed slightly. "Well, not *saleable*, exactly," he admitted. "If there was something in there to interest Big Pharma the suits would have been round your house waving contracts at your Dad before the results ever got to the school. I was thinking of some talent or aptitude that you haven't discovered yet—something that might conceivably make you money later in life, or at least guide you into the kind of job where you'd be most likely to be happy."

"Did you talk to Helena?" Janice asked, abruptly changing the subject.

"No—but I did notice that she came back to class looking as sick as a dog. Whatever it is, she probably won't want me to know about it. She seems to be confiding in Crispin, though. He'll probably tell you, if you don't want to ask her yourself."

"Why wouldn't I want to ask her myself?" Janice said, resentful of his obvious awareness of the problems she and Helena were having.

Tim shrugged. "I don't know," he said, defensively. "Why did you ask me?"

"Because you're supposed to have all the answers," she snapped.

"No I'm not," he replied. "I'm just the one with the bootleg software that knows a few more answers and a lot more possibilities than the stuff the government issues to schools."

"Could it help Helena?"

"It might—but probably not, if she's got a seriously bad gene. If that's the case, she can probably get the best advice that's currently available from her family doctor's consultancy software. My stuff deals with issues that are less clear cut. You really ought to let me run it over your chip—you already know that there's no bad news, so why not go fishing?"

"I'll let you know," Janice said—not thinking of herself, but of Helena. If the news Helena had received from Mrs. Charters was bad, she might be far more in need of crumbs of comfort than Janice was, and now that the two of them were speaking again, after a fashion....

Unfortunately, Janice didn't get a chance to talk to Helena before the bell went again. Once they were back in class Helena and Crispin were separated, but so were Janice and Helena, and because the afternoon timetable had Maths followed by Geography the opportunities for private conversation that might have been thrown up by Eng Lit or Lab Science didn't arise.

When the final bell went Janice tried to catch up with Helena, who was sitting closer to the classroom door, but she was intercepted—not by Tim, which would have been annoying, but by Crispin, which shouldn't have been.

"Sorry," Janice said, as her haste nearly caused her to barge into him.

"That's okay," Crispin assured her, effortlessly continuing to impede her progress. "I saw you talking to Timmy the geek in the lunch-break."

"So what?" Janice said, trying to suppress the blush that she felt rising to her cheeks.

"So nothing—except rumor has it that he's got hold of some bootleg genome-analysis software."

That made Janice start slightly. "So he says," she admitted, warily. "Why do you want to know?"

"I just thought, maybe, that it might be worth...well, running some stuff through it." His hesitancy spoke volumes. If he'd been talking about his own profile, he wouldn't have been so reluctant—and he wouldn't have been trying to recruit Janice to his cause.

"Why don't you ask him?" Janice said, exaggerating the reasonableness of the question as much as she could.

"Because he might say no to me," Crispin was forced to admit. "He wouldn't say no to you, even if you wanted to include a few other people."

"A *few* other people?" Janice queried. "How many exactly?"

"Well, me...and Helena."

Janice could see that Crispin was well aware of the sensitivity of the topic. He obviously knew that mentioning Helena in the same sentence as himself might have an adverse effect on Janice—which meant that he'd either deduced, or had been told, that Janice was jealous of Helena's increasingly close relationship with him. The ramifications of that awareness were not something Janice wanted to contemplate at present, though.

"Helena could ask him herself," Janice pointed out. "Tim wouldn't say no to Helena—if she asked nicely, that is."

It was Crispin's turn to blush slightly—which amazed Janice, who hadn't known that he had it in him. "Well, that's another thing," he said, after a pause. "I thought maybe you could...have a word with her too. I mean, you're still friends, right? You could do that."

"You're worried about her?" Janice said, thinking that at least it demonstrated his capacity for sensitivity, even if it would have been infinitely preferable had the sensitivity in question been exercised on her behalf rather than that of her ex-best friend.

"Yes, I am," he said, in a tone so unconvincing that Janice immediately realized that she might have jumped to the wrong conclusion. What had probably happened was that Helena had refused to tell him the reason for her black mood, in spite of his solicitous questioning. Crispin didn't want Helena to keep secrets from him, and he thought that if Janice could talk her into letting Tim Sillington run her chip through his bootleg software while he was present, Helena would no longer have a secret to keep.

In any case, Janice thought, Crispin's request offered an opportunity of sorts. It offered an opportunity for her and Helena to stand in Crispin's sights simultaneously, in a context in which there would be more than mere beauty at stake. It offered an opportunity to move the contest into a new phase. If Tim's compatibility-analysis software happened to reveal that Crispin was more compatible, genetically speaking, with Janice than with Helena, then maybe he might look at her in a kindlier light. There was, of course, no guarantee that it would, but Janice was prepared to take the risk.

"I suppose I could have a word with her," Janice said, refraining from mentioning the fact that she had been on her way to do exactly that when Crispin had got in her way. "If you want me to, that is."

"If you wouldn't mind. She'd listen to you, I think. She's a bit upset, though. You'll have to be careful—but you're good at that. Better than me, at any rate."

The compliment was so weak that it was difficult to be grateful for it, but Janice tried. "Well," she said, "we haven't been getting along very well for the last few weeks, but if you want me to make the effort...." She trailed off, as artfully as she could.

"Please," Crispin said: a word that would have been worth its weight in diamonds or rubies, if it had had any weight at all.

"Okay," she said. "Leave it to me—and I'll take care of Tim, too. I'll fix it. Would tomorrow suit you?"

"Fine," he assured her, even contriving to add: "Thanks, Janice."

That was, Janice thought, as she was finally able to make her way to the school gates, the nicest combination of syllables she had heard all day.

* * * * * * *

Janice let an hour elapse after she'd eaten her dinner before she went round to Helena's house. By that time, she figured, Helena would be feeling so much pressure from her parents' relentless solicitude that she would be glad of a legitimate excuse to escape. She didn't ring first, or even text, in case Helena might react too quickly with an automatic rebuff. Her physical presence would be much harder to repel, especially since Helena probably hadn't told her parents that she'd fallen out with Janice.

The plan worked perfectly. Helena's mother answered the door and ushered Janice in with an alacrity that suggested that she too was feeling the strain. After a few moments of carefully-concealed resentful seething, Helena came to the inevitable conclusion that Janice represented the lesser of two inconveniences, and took her upstairs to the sanctuary of her bedroom.

"What do you want?" Helena demanded, as soon as the door was safely shut.

"I was worried about you," Janice said. "I knew when you came out of old Charters' office that you'd been hit hard, but I had to go face the music myself. I couldn't get near to you at lunch time, or in Maths or Geography, so I came over when I could. You don't have to tell me what it is—I just wanted you to know that I'm here for you. Why weren't your parents notified in advance? They must have got your results before the school did."

The last question was cleverly calculated to give Helena an opportunity to sound off, which Janice judged—correctly—that she would be unable to resist. "They would have," Helena said, disgustedly, "if they'd bothered open the email. Because he works for the Department of the Environment, Dad gets hundreds of official emails, almost all of them flagged urgent, and he's always at least a week behind. Mum, of course, leaves all that sort of stuff to him. She makes him deal with all her official email as well as his own, no matter how loudly he complains about not having the time or how long he leaves it in the hope of forcing her to do it herself. Not that it was news to them, anyway—Mum's no genealogy freak, but she's no stranger to gossip. She knew more than enough about her family history to figure out what the profile was going to tell us, but she never felt the need to mention it to me. Didn't want to worry me before there was any need, she says."

"It's a vulnerability to some kind of hereditary disease, then? How bad?"

"Bad enough." Helena hesitated then, but eventually shrugged her shoulders and elaborated. "Aggressive cancer, initially prone to affect the liver, pancreas or various endocrine glands. Could be triggered any time, by random mutation, liable to spread too rapidly for effective containment by surgery or radiation therapy and often unresponsive to chemo. Given the levels of air pollution in the Thames valley, the degraded state of the local aquifers and the ever-present terrorist threat to hit London with a dirty bomb, I'm not in an ideal situation to avoid the hit, but moving to the Outer Hebrides doesn't seem to be a viable option. That's not the only nasty one, but the others aren't as urgent."

"I'm sorry." It was all Janice could think of to say.

"I told you, didn't I? When we were in the queue. I told you I didn't need to know about any guns that were pointed at my head. Why suffer the fear as well as the risk? It's not as if they can do anything to help, no matter what they say about the irresistible march of bloody progress."

In the face of news this bad, Janice knew, there was little point in suggesting that Tim Sillington's bootleg software might be able to point Helena in the direction of some previously-unsuspected aptitude or talent. On the other hand, given that Tim was in a position to acquire bootleg software, by whatever means, he might be able to locate information specifically relevant to Helena's problem that wasn't publicly available as yet. Before that possibility was broached, however, there were other enquiries that needed to be made.

"Can your Dad help?" she asked, tentatively. "I mean, he does work for the government."

"Half the people in our road work for the government," Helena said, scornfully. "It doesn't give us any extra privileges. Just because he's in the Department of the Environment, it doesn't mean that he can clean up the air or the water. We buy our filters at Sava-Centre and monitor our meat and fish just like everybody else. I suppose I could wear an oxygen mask, like people do in Tokyo, but...well, it wouldn't be me, would it?"

"You'd be just as good-looking under the mask," Janice pointed out.

"But nobody would know," Helena countered. "Who'd take the trouble to remember?"

Janice inferred that the last bitter remark had been voiced with Crispin Slipman in mind. "Who else have you told?" she asked, delicately.

"No one, yet," was the answer. "I wanted to have it out with Dad first. But there's no point in keeping it a secret, is there? You don't have to swear that you won't pass it on. Tell whoever you like—even Timmy the geek."

Janice started slightly, but she realized almost immediately that she shouldn't have been surprised. The thought that had occurred to Crispin had also occurred to Helena. Tim Sillington's image problems were undergoing a miraculous transformation. Although it had never occurred to her that the compilation of the National Genetic Database would have a profound effect on the way that certain designated outsiders would be regarded by their peers, the logic of the situation was perfectly clear in hindsight. Helena had already had all the advice that Mrs. Charters could give her, and had found no comfort therein. Now she wanted a second opinion—and where else could she look for one?

"It might be worth mentioning it to Tim," Janice said, carefully. "So far as I know, his bootleg software is all pseudo-astrological stuff calculating the number of genes people have in common with each other, and with the rich and famous, but if anyone can get more...."

"Would you ask him for me, Jan? He doesn't like me."

"Only because he's frightened of you."

Helena seemed honestly amazed by this allegation. "Me?" she said. "It's the lads who rough him up." That was true, Janice knew. It wasn't the year elevens who ran the school protection racket who were the worst problem, but the ones way down the pecking order, who needed easy victims on whom to take out their resentment.

"It doesn't matter," Janice said. "I'll ask him if we can go round to his tomorrow night. Would that be okay?"

"The sooner the better," Helena answered. "Not that I'm expecting anything, mind. I just want...an insider's eye."

Janice wasn't at all sure that Tim's eye qualified as an insider's, given that he couldn't even hack into the police databases, but she knew what Helena was getting at. While the bad news was all coming at her from above it must seem like a hailstorm from Hell. She wasn't expecting a secure shelter from the storm—she just wanted to discover that people in her position weren't entirely powerless.

"I'll fix it," Janice promised. "After all, what's a best friend for?"

All she got in reply to that was a nod, but it was a precious nod, because it signified that they were once again best friends, in spite of everything.

* * * * * * *

The following afternoon, as soon as they'd changed out of their school uniforms, Janice and Helena met up and went round to Tim Sillington's house together. Helena was astonished, and not entirely delighted, to find Crispin, still in his uniform, already there, but she obviously remembered what she'd said to Janice about there being no point in keeping such matters secret.

"Got your chips?" Crispin asked, as they all climbed the stairs to Tim's bedroom. "I've got mine." He held it up, by way of demonstrating that they were all in this together. He still didn't know any details about Helena's bad gene—Helena hadn't told him, and Janice hadn't felt free to break the news.

Janice and Helena both had their chips, but neither of them wanted to brandish them around like trophies. Janice figured that there would be time to bring hers out of hiding when it was her turn to surrender it to the occult powers of Tim's magic software.

Janice hadn't been in Tim's room before. She was expecting a wizard's lair crammed to bursting with arcane equipment. She was slightly disappointed when she discovered that Tim only had one PC, which didn't look significantly different from hers, and a CD tower that was small enough to stand on his desk. He didn't have any posters on his walls advertising obscure death metal bands or prints of images from the Hubble Space Telescope. His duvet matched his pale blue wallpaper and his chest of drawers was from Ikea.

Crispin was slightly more impressed by the PC, obviously having noticed features that were invisible to Janice—or, at least, being prepared to pretend that he had. Crispin grabbed the second chair at the desk, leaving Janice and Helena no alternative but to perch side by side on the bed, leaving just enough gap to signify that their re-alliance was still a trifle fragile.

"My data's already loaded," Tim said. "Who's next? Janice?"

Janice blushed, but it would have been too cowardly to insist that Crispin or Helena put their data on the line before she surrendered hers. She handed over the chip.

"It'll only take a minute," Tim said. Janice guessed that it wasn't strictly necessary for Tim to watch the screen so intently while the data loaded, but it gave him a good excuse to avoid eye contact. Crispin, on the other hand, didn't know where to look, especially as Helena was very obviously not looking at him.

"This is a mistake," Helena said, although she made no move to leave.

"It can't hurt," Janice assured her, "and it might help."

"Janice is right," Crispin was quick too say—but then it was his turn to hand over the chip containing the confidential record of his biological make-up. He couldn't help hesitating, and looking a little anxious. The hint of vulnerability, Janice thought, made him look even more attractive.

Finally, Helena handed over her official death-sentence.

"The first thing I'm going to do," Tim said, "is run the four profiles through correlation programs that will search for significant similarities between their make-up and the characteristic profiles of successful individuals in various professions, vocations and so on. Don't read too much into this—it's simply not the case that all professional footballers or all painters have particular arrays of genes in common. There are statistically significant correlations within the groups, though, so if you have a significant correlation with the group mean, it might be worth searching for more specific indications of a genuine aptitude. On the other hand, it's not like dietary or medical sensitivity, where the revelation of a variant gene can indicate an urgent need to avoid certain kinds of food or prescription drugs. If you happen to hate Maths, you're probably better off ignoring any slight correlation your profile may have with the population of physicists."

Janice suspected that the last remark might be a slight dig at her, but it was slight enough to let it pass. Even so, she said: "And what does your personal profile correlate with?" she asked. "The population of great computer scientists?"

"Actually, no," Tim answered, without any conspicuous embarrassment. "The highest correlation coefficients I had relative to employment groups were with actuaries and public service accountants."

"What's an actuary?" Janice had to ask.

"Someone who calculates the likelihood of people dying, for the benefit of life insurance companies," Crispin supplied.

"Near enough," Tim admitted, moving swiftly on to add: "Crispin, your highest correlation coefficients are with stockbrokers and health service administrators."

"That can't be right," Crispin objected. "What do stockbrokers and health service administrators have in common?"

"The management of risk," Tim replied, promptly. "What the software's implying is that you're a good gambler—good in the sense that you're naturally inclined to play safe without being overcautious."

"They also do their risk-taking second hand," Helena put in, causing everyone to look at her in surprise. "They play with other people's money and other people's lives."

"Janice," Tim was quick to say, as more data flickered on to his screen, "your highest correlation coefficient by far is with...mothers."

"Mothers?" Janice echoed, unable to keep the resentment out of her voice. "You're telling me that your bloody computer thinks that I'm likely to fall pregnant before I leave school!"

"No!" Tim was quick to retort. "That's *not* what it means. The correlation isn't with women who have a lot of children, or who start early, but with women who qualify as *successful* mothers. It doesn't necessarily mean that they don't do anything else."

Janice took due note of the deadly *necessarily*, and also of the fact that whereas Tim and Crispin had both been given alternatives, she had been gifted—or cursed—with one correlation coefficient far higher than all the rest.

"Hang on," Tim added. "I think I get it. Did Mrs. Charters tell you that you had an unusually clean profile, health-wise?"

"Yes," Janice confirmed. "*As close to problem-free as any mere human is likely to get*, I think she said."

"Well then, it's not surprising that the program throws up a correlation with successful mothers. Gene-defects have a considerable effect on the chances of successful motherhood, not only because they're heritable, having an inevitable knock-on effect on the life-chances of any children the person in question might have, but because many of them have a direct effect on the chances of falling pregnant or of successfully carrying an embryo to term. It's a negative correlation rather than a positive one—it doesn't mean that you have something in your nature that inclines you towards a career as a mother; it just means that you have nothing in your nature that would get in the way—nothing that would prevent you from being a successful mother."

"But that's good, right?" Crispin said. "It must also mean that there's nothing that would get in the way of pretty much anything else Janice wanted to do."

"Except that there isn't any other correlation to tell me what I might have a positive aptitude *for*," Janice pointed out, bitterly. "I might do anything, but there's nothing I'd be particularly good at—so nothing shows up in the numbers except the fact that I'm slightly less likely to fuck my kids up than most mums."

"No Janice, Crispin's right," Tim said. "It's a good thing not to have your choices narrowed down—and it's certainly not a bad thing to know that if you do want kids, you've a far better chance than most of doing a good job."

"What about me?" Helena's plaintive whisper was only just audible, but it reminded all of them of the real reason they were there. The relevant data was already on the screen, waiting for Tim to read it.

"Your top two correlations are with singers and models."

"That doesn't make sense," Helena retorted. "I can hold a tune, but I certainly don't have the kind of voice that's trainable for grand opera, and there's no way I'll ever be tall enough to be a model, even if I become pretty enough. It has to be wrong.

"I'll refine the categories," Tim said. After a moment's pause, he added: "The correlation is with singers of popular music rather than opera, and you're right about the height thing cutting down the correlation with catwalk models—but that's not the only kind of modeling the software takes no account."

"Are you telling me that I'm supposed to get my boobs done and pose for porn mags?" Helena demanded, her anxiety demanding momentarily into anger.

"No!" Tim replied, blushing deep red. "But height isn't that restrictive as a limiting factor, given that you do have such a fabulous face. There are...." He stopped, presumably because he had seen the color drain out of Helena's allegedly-fabulous face.

"What is it?" Janice asked, moving to eliminate the distance they'd carefully set between themselves and put a supportive arm around her friend.

"What Tim said about you," Helena muttered.

"I don't understand," Janice told her. "Why would anything Tim said about me make you frightened?"

"About it being a negative correlation, not a positive one," Helena said. "I don't have those correlations because I've got any positive aptitude for music or modeling—it's just that those are categories packed full of people who make their mark on the world before they're thirty...in which even people who die young can be successful."

Tim opened his mouth as if to say "no" again, but then he shut it with the word unvoiced.

Crispin also opened his mouth, presumably to ask what Helena meant about dying young, but he too choked the words back. Janice read in his features that he now had a pretty good idea of what it was the Helena had been so reluctant to tell him.

Tim swallowed, and started again. "That's the next step," he said. "We need to look more closely at the correlates of the specific kind of cancer to which you're vulnerable. We need to look at recent medical research, to see what specific treatments might be in development. This is the twenty-first century—every decade brings out new technologies, and aggressive cancers are right up in the front rank as key targets for all manner of magic bullets. Given your age, and the fact that they've only just put the NGD together, you're a prime candidate for clinical trials of preventative measures. If I can identify something that you can actually take to your doctor...or even direct to source...."

He stopped, because Helena had stood up. "This was a really bad idea," she said. "I need to go home." She looked at Crispin, but Crispin didn't look back at her, let alone leap up and offer to escort her home. Janice could see plainly enough that it wasn't because he wanted Helena to sit down again, so that Tim could continue trawling for fugitive atoms of good news. Crispin had already found out what he had come here to determine, when Tim pronounced the magic word "cancer". Janice could see that Helena had just been severely devalued in Crispin's eyes. He was, after all, able to look beyond mere appearances, to search for inner beauty behind a mask—or maybe he just had some sort of phobia associated with the mention or the idea of cancer.

Janice was abruptly put in mind of the old saw about being careful what you wished for, in case you got it. Crispin's newly-revealed immunity to superficial charms didn't seem in the least like maturity or sound romantic judgment. It seemed like another kind of meanness.

Crispin looked at Janice then. He looked at her in a reappraising fashion, which suggested that he'd meant what he'd said about the correlation Tim's bootleg software had made for her being a good thing. Janice knew that Crispin wasn't anywhere near the point in life where he might begin thinking about finding an appropriate mother for his children, but she knew now—thanks to Tim's software—that he was a person with an inbuilt natural inclination to play safe in managing risk, to respect negative correlations and to go with the choice that was least likely to lead to trouble.

"Come on, Helena," Janice said. "You're right—this was a bad idea. I'll walk you home."

"I don't *need* you to walk me home!" Helena spat, venomously. "I'm *not sick*."

"No," Janice said, softly. "Of course you're not sick, and of course you don't need anyone to walk you home—but I'd like to do it, if it's okay with you. We're friends, remember."

"You're not my mother," Helena riposted, much more weakly. "I can look after myself."

All Janice said in reply to that was: "Let's go." And they went, together.

* * * * * * *

Later, when Helena was safely ensconced with her mother and father for a resumption of their deep and meaningful family discussion, Janice went back to Tim's. She expected Crispin to be long gone, and she wasn't disappointed.

"You were right," she said to Tim, as she sat back down on the bed while he resumed his seat at the PC. "The NGD *will* change everything. It *does* change everything."

"I've been checking the stuff I said I'd check," Tim told her. "Helena's situation really isn't as bleak as it might seem. There are several research programs targeting the kind of cancer that she's vulnerable to, and now that she can be clearly identified as a prime candidate for clinical trials, she can probably take her pick of them. I'm not the right person to advise her on which one to go with, but her GP's consultancy software will help her make an informed decision."

"Did Crispin say anything to you after we left about *his* informed decision?" Janice asked.

"You mean, did he say anything about you?" Tim asked, warily.

"No," Janice said. "I mean, did he say anything about dumping Helena?"

"Isn't it the same thing?" Tim said, warily. "No, he didn't say anything—but I think you might have edged ahead in the competition."

"There isn't a competition any more," Janice told him. "Crispin Slipman is no longer a glittering prize—not so far as I'm concerned, at any rate." She had to pause then, and think about whether that comment made her just as shallow as Crispin, willing to chop and change her affections on the basis of information that really shouldn't have been relevant to matters of the heart. She decided, after a few moments thought, that it didn't.

"I'm sorry," Tim said. "About what the program said about you, I mean. It's silly—just a statistical thing. It doesn't mean anything."

"And what about the other programs?" Janice asked. "The compatibility programs that are supposed to be so much better than astrological calculations of who's best suited to be with whom?"

Tim blushed, and hesitated—but then he bit the bullet, like the hero he was, and told the truth. "Crispin had a far higher compatibility rating with you and Helena than I have with either of you," he admitted. "Your compatibility rating with Crispin's is higher than Helena's. Helena's compatibility rating with me is slightly higher than yours, although both would probably count as contraindications."

"All nonsense, then," Janice said.

"Not necessarily," Tim said, bravely. "It just means that my fancying you is the absurdity that you've always considered it to be. You were probably right when you said that the only reason I don't fancy Helena more is that I'm afraid of her."

Janice didn't bother to point out that what she'd actually said had been far more moderate in its implications. "Well," she said, "now you know that all I'm good for is motherhood, you can safely cross me off your list of desirable partners. Not clever enough, and not a single latent talent. You'll get over it easily enough—and you have your machine to help you find someone more suitable."

"The correlation coefficient doesn't mean that motherhood is the only thing you'd be good at," he said. "That's a fundamental misunderstanding of the nature of correlation. Your result *doesn't* signify that you don't have any latent talents—it signifies that you're not restricted to any particular narrow range of talents. You don't have any innate barriers to achievement. It's not that you can do *anything*, exactly, but that anything you actually want to do—anything you're prepared to take an interest in and work at—you *will* be able to do. Than gives you a big advantage over me...and Crispin too. I know you think that there are all sorts of things you're no good at, but what your profile says is that it's mostly just a matter of not liking things you haven't tried. You ought to stop making excuses, Jan, and accept that you really can be whatever you want to be."

"No I can't," Janice retorted. "You just mean that whenever I do decide what I can and might do, I won't run into any limitations specifically imposed by my genes. And that my kids will be fit and healthy, provided that I take a good look at my potential partner's profile. It's all negative, remember—all I'm guaranteed is a few extra layers of insulation against the probability that things will go disastrously wrong. They won't stop me being hit by a car if I'm not careful enough crossing the road. See—I'm not quite as stupid as I look."

"I never thought you were stupid," Tim told her.

"Even when I was infatuated with Crispin?"

"It wasn't a relevant issue. Do you want to know what else I found out about you by running your profile through my software?"

Janice as tempted to say that he'd done enough damage already, but she knew that it wouldn't be fair and that he'd be hurt.

"Why not?" she said. "I might as well take advantage of being first in the queue." She was fairly certain by now that there would be a queue. There would undoubtedly be a more than a few people who wouldn't want their profiles checked by Timmy the geek, even if he did have the best collection of bootleg software in year nine. Curiosity was a powerful motivator, though—and most people, however absurd it might be, didn't think that having their genes analyzed was an invasion of their privacy in the same way that having their emails read would have been. As Tim's reputation grew, on the basis of what the queue reported back, it would doubtless extend beyond year nine, perhaps even as far as the bullies who sometimes gave him a hard time—at which point, he might no longer need to hack into the police computer to begin turning the tables on them.

"You can sit here if you like," Tim said, indicating the other chair. "You can pull it up as close as you like—to the screen, I mean."

"I know what you mean," she said. She moved to the chair, and pulled it as close to the screen as she liked.

She knew that everything, in one sense, was still the same. Everyone was exactly the same person as he or she had been before the NGD had delivered its news regarding exactly what they were made of. And yet, everything had changed. Knowledge changed the way that everyone looked at the world, and once everyone got used to the idea, everyone would want to look harder, because they wouldn't be able to afford to look away.

"Helena will be all right, you know," Tim murmured, when Janice's face was close enough to his to facilitate whispered speech. "At least, she has a far better chance of being all right *now* than she would ever have had in the olden days."

"I know," Janice said. "The same goes for me—and for you. Crispin too."

"He'll probably ask you out, you know," Tim said. "Even if he doesn't seem quite the glittering prize he once did...my mother's always saying that you're only young once, and that she wishes she'd taken more advantage of it."

"My mum says the same thing," Janice told him. "It's probably true—but knowing it doesn't make things *that* much easier, does it?"

"Right," Tim said, as his fingers fluttered on the keyboard and the image on the PC screen shifted yet again. "Now, pay attention—this is as close as anyone can get, just now, to the innermost essence of the ultimate you...."

Janice already knew that wasn't true. She already knew that it couldn't be true, because the ultimate her was something that hadn't been made yet, and wouldn't be complete for some time. On the other hand, she also knew that if she were to make a good job of making the ultimate her, she'd probably need all the help she could get, not just from Tim's computer but from Tim himself—not to mention Crispin the prudent gambler, and her beautiful best friend Helena, and everybody else.

Bit by bit, the news began to unfold on the screen.

BURNED OUT

When Carmichael went to walk through the ruins with Burke, the forensic scientist from Ashton, he told Sergeant Andrews to stay with the jeep. Andrews obeyed, but didn't relax; he used the binoculars to scan the grounds carefully. Andrews was a scrupulous man, who took his missions seriously. He didn't like being on bodyguard duty, but he wasn't about to get lazy—or to let his men get lazy—simply because he didn't like his job. Carmichael knew he ought to admire that, but in fact it made him feel claustrophobic.

There wasn't much to see, with or without the aid of binoculars. The establishment had been obliterated; the roofs had caved in and the walls had collapsed. Everything that had been inside the buildings was reduced to powdery ash. The institute had been scientifically torched by people who intended to leave no traces; the operation had been carefully planned and executed.

"This is where they found the human remains," Burke told him, pointing to a mess of comprehensively-sifted rubble. "The animal stuff too. Even the bones were all-but-gone. We tried to use the teeth to check IDs, but it wasn't easy. We think there were five human bodies and two others, but I can't even be sure that the others were chimpanzees. We have two premolars and a molar with fillings that match the dental records you sent through for Abel, but I can't be absolutely sure that Franklin was one of the others. I could only confirm the identity of one of the local men with any real confidence—and we have seven names listed as missing. We may never know exactly who died and who didn't."

And by the same token, Carmichael thought, *we may never know exactly which of them should be numbered among the perpetrators, and which among the victims.*

He felt numb inside at the thought of the waste, and the stupidity of it. Even if the lab had been doing the kind of work the anti-biotech extremists thought it was—even if its sealed chambers had been brim-full of armaments for use in the ongoing plague war—this would have been a meaningless act, a gesture of blind rage. Always assuming that it had been members of the anti-biotech brigade who were behind the attack. That hadn't been proved yet. Nothing had been proved yet.

"As you can see," Burke went on, defensively, "We don't have a lot to go on. The police really are doing the best they can."

"I don't doubt it," said Carmichael, mechanically.

"I can show you all our lab data—and what's left of the gleanings. You can check it over—take stuff back east with you if you want to, for genetic fingerprinting or whatever. If you can get a result, you're a better man than I am."

"No one doubts your competence, Dr. Burke," said Carmichael, patiently, "And no one doubts the enthusiasm of the local police. I'm just here to compile a report. You know how things are."

Burke nodded to indicate that he did, but Carmichael knew it was a lie. Nobody knew how things were—not any more. Things were coming apart at the seams, and you couldn't rely on any of the old routines.

They went back to the jeep and got aboard.

Andrews waited until all his men were loaded into the truck before telling the driver of the jeep to move off. He wanted the vehicles in close conformation while they rode into the town, just in case the road was booby-trapped and the territory was lousy with enemy agents.

In fact the road was empty, and the surrounding countryside was utterly desolate.

"Did you know Abel and Franklin personally?" Burke asked, after they'd gone a mile or so in silence.

"Not very well," said Carmichael. "I worked in the same building as Abel once, but we were on different projects. I met Franklin at a conference back in 2017." He spoke disinterestedly. He wouldn't have been grieving for the two biotechnologists even if he'd known them better. With thirty million dead in little over six months it was difficult to feel grief for individuals any more. Nor was the project any great loss, from the point of view of the ongoing war effort. It had been a speculative thing, funded in days when priorities were different. That was one reason why the place had been understaffed and underprotected—easy meat for the arsonists.

"I don't think it was local people who did it," said Burke, uneasily. "They're a little crazy, with the war and everything, and there were some wild rumors circulating about what the labs might have been cooking up, but that wasn't the work of some over-excited mob. It was an expert job."

Carmichael nodded. Burke was probably right—except that it didn't quite add up. Some of the anti-biotech groups were extremely committed, and not without expertise in matters of sabotage, but the most highly-organized also had the best intelligence, and they'd have taken the trouble to make sure that Abel's establishment really was doing plague war work before sending in commandos. Local people, on the other hand, were much more vulnerable to crazy talk about the war not being a real war at all, but simply a series of escapes from the government's own labs, possibly engineered by mad eugenicists trying to slim down the econim population. All nonsense, of course—but way out here, the people had always regarded the distant federal government as a false friend and potential oppressor.

The plague war wasn't the kind of catastrophe that enhanced nationwide social solidarity; the common enemy was too diffuse, too uncertain. Instead, communities were drawing in upon themselves as people lost faith in Washington, in the Union. As it slowly came to seem that the whole world was collapsing, backwoods folk became increasingly desperate to preserve the land between the horizons they could see—and many of them refused to think beyond. The survivalist ethos had taken off in a big way. These were crazy times.

Plague war wasn't like the kind of wars people had been taught about in history; there was no army for the young men to join, no evident enemy to fight. The only responses people had to the deaths of their kids and their cousins were containment and cauterization: trying to burn out the plague germs wherever they appeared, and limiting their spread. It was easy enough for some of them to turn their fearful eyes and frustrated ire upon the government's own biotech research establishments, figuring them as part of the problem rather than part of the solution. It was easy, but it was dead wrong; if there was hope for the future, Carmichael thought, it was contained in the labs that were working flat out to find effective defenses for use against viral and bacteriological terrorism.

"On the other hand," Burke said, "the townspeople never wanted the lab here. A lot of people thought that it was forced on them. Nobody ever took the trouble to tell them that the people there weren't engineering viral weapons. Nobody told them much of anything, except maybe to mind their own business while the world went to hell."

"People like Abel and Franklin are trying to save the world," said Carmichael, disgustedly. "Whether genetic engineers are somehow to blame for the war or not, we're the only people who can possibly rescue the human race from its effects. The men that torched that establishment aren't just treasonous, they're insane."

"Hell," said Burke, "I know that. It's just that...."

Sergeant Andrews was looking the other away, pointedly ignoring the conversation—but he was listening. Carmichael knew that Andrews was neither treasonous nor insane, but he also knew that the sergeant didn't feel any particular loyalty to the men who'd made the weapons of the ongoing war. He was a soldier; he'd been trained to fight enemy soldiers from armored positions, firing good honest bullets. Being shoved into the front line of this kind of conflict must have seemed to him a kind of betrayal, and, like everyone else, he was bound to be tempted by the simplistic explanation that it was all the fault of the scientists whose work had made such warfare possible.

The streets of the town were almost as deserted as the road from the burnt-out lab-complex. Half the houses were empty, already falling derelict. The people who remained stayed indoors as much as possible, fearing that every breath of wind might be contaminated.

Even so, life went on: work, commerce, social life. The shops had little enough in the way of goods, but they were open for business and gossip, and people were doing their damnedest to keep things going, to maintain their routines.

Burke guided them to a hotel in the centre of town, which still had a full staff, and they booked in. The sergeant sent two of his men out on a foraging expedition to buy in supplies.

"I'd to speak to someone who worked at the labs," Carmichael told Burke. "If you could find me someone who can tell me what was going on in the weeks before the attack, I'd be grateful. I need to get a better idea of the stage the experiments were at. Abel was a little behind with his reports, thanks to the war and being so short-staffed."

"I'll ask," said Burke, "but it might not be easy. The survivors aren't exactly in hiding, but some of them became very shy about what they did long before the fire. You know how it is, with the rumors and all."

Carmichael knew how it was. Even cleaners and gate-guards who worked in biotech installations had become shy since the war began—and their reluctance to talk to strangers would amplify that all-round reticence. Everybody in Ashton would have been happier if he and his guardian angels had not come.

"Do what you can," he said, with a sigh. "The quicker I can find something substantial to put in my report, the quicker I can get away."

* * * * * * *

After he'd eaten supper with Andrews and the men Carmichael took a flask of boiled water up to his room. He'd managed to locate a working TV set but the official broadcasts were all routine and both the ents channels were reeling out ancient tapes he'd seen way back when. It didn't take long to write up the day's notes, and when he'd finished he loaded a chip into his bookplate and settled down to read. He was just getting absorbed when somebody tapped on the pane of the curtained window.

He was startled at first—he was on the third floor—but he realized that someone must have climbed up the fire escape. He went to the window and opened it. It was raining outside and the woman who was waiting was pretty wet. She was in her mid-twenties, with short-cropped brown hair and steel-rimmed glasses. When he let her in she took time out to clean the raindrops off the lenses.

"Dr. Carmichael?" she said.

"Yes."

"I heard—indirectly—that there were army personnel here, asking around for anyone who might know what had happened up at the labs. Your name was mentioned." She slipped off her wet coat and draped it over the back of the room's only armchair.

"What's wrong with the door?" he asked, thinking: *I misjudged you, Dr. Burke—I didn't expect results this fast, if at all.*

"I'd rather keep this between the two of us. There's some bad feeling about the labs and what happened up there, and I wouldn't want people to think I'd come here to finger anyone."

"Have you come here to finger someone?" he asked, wonderingly.

"Of course not. I don't know who torched the place or why—I was off shift, tucked up in my bed. If I hadn't been, I guess I might be dead. But there are some crazies around, you know, and there was a certain amount of talk about what certain people thought the good people of Ashton ought to do about the germ-factory on the ridge, and some of those certain people now want to keep a low profile. I told anyone who'd listen that the germ-factory stuff was all nonsense, of course, but no one believed me. Anyhow, I didn't want it noised around that I came running to you the moment you breezed into town."

He shrugged. "I'm not here to chase the people who did it. I'm just here to make a report about the work Abel and Franklin were doing—not that it's going to be easy. Whoever wiped out the labs did a very thorough job."

"Abel and Frankin are both dead, Dr. Carmichael. So are the apes. They were in quarantine—they never got out."

"What apes?" he asked, carefully.

"You must know about the apes" she replied, equally carefully. She squinted at him through the polished lenses, and it was obvious that they were both wondering how much the other knew.

"To tell you the truth," he said, "Abel's reports may have been more than a little out of date. He wasn't always punctilious about making official reports on experiments in progress...I don't suppose anyone is. We all like to have the results in before we report to our masters. His working notes must have gone up in smoke, unless he kept back-up discs at some distant location. The only chance we have of finding out exactly where he was up to is to find someone who was actively involved—however limited their role might have been. Exactly what did you do, Miss...?"

"Vollman—Lucy Vollman," she said. He recognized the name from the list he'd scanned, but couldn't connect it to a role. "I started out as an equipment orderly," she added. "Bottle-washer, animal-handler and all-purpose spare pair of hands. They promoted me to technician when the people they had were recalled to war work." She must have seen his face fall, because she went on: "That doesn't mean that I'm a moron, Dr. Carmichael. I was more than half way to working my way up to technician anyway. It can be done, you know."

He knew that she was right. Nobody tolerated fools in high-security labs. No matter how menial the tasks to be carried out, they had to be done efficiently and with intelligence. It was sometimes better to recruit technicians who knew the lab and its routines than youths straight out of school—and it was certainly useful to be able to keep junior staff on their toes with that kind of dangling carrot.

"How much do you know?" he asked, sitting down on the bed and pointing to the armchair. She moved her wet coat to one side and sat down in it.

"Abel never began any human experiments," she said, without further procrastination. "He was probably disappointed with the way things went with the apes. There were just the three chimps—or what had been chimps. They're all dead, reduced to ashes. You do know about the chimps, don't you?"

Carmichael nodded.

"That's it, then," she said. "If you know about the chimps there's nothing more to tell." But she didn't make a move to get up out of the chair.

"I need to know exactly what happened to the chimps," he said, as she had known that he would, "as accurately as you can tell me. I'd also like to know who else knew."

"I don't think anyone burned down the labs because of the apes," she said, mistaking the reason for his last inquiry. "Something may have slipped out about them—garbled, of course—but it wasn't the kind of horror story to excite firebugs. You know there are rumors flying around that the plagues were actually cooked up by the government, don't you? People whisper it back and forth that Washington is using them to wipe out the blacks, or the Hispanics, or miscellaneous trailer trash, while the rich hide out in their bunkers. They talk about international conspiracies of the rich against the poor, about human culling...all kinds of crazy stuff."

"It's all lies," said Carmichael, "but...."

"That's what most people say," she interrupted, "but what the crazies reply is, *how would you know?* And we wouldn't, would we? If things like that were going on, who'd tell the likes of us. Me, anyway. Do you think Dr. Abel kept back-up discs somewhere else, Dr. Carmichael? Would they be any use if there were any, now that you know that the next phase never got off the ground?"

"Probably not," he said, off-handedly, wondering who was supposed to be interrogating whom. "Who else do you know who might have information? Any information at all."

"Less than half a dozen," she said. "Only two who worked on the inside. Nobody knows any more than I do, Dr. Carmichael. I don't think any of the others is going to come tapping on your window, and I'm sure that none of them is going to march into the lobby and ask your soldier-boys where to find you."

"I need you to tell me where I can find the others," he said, stubbornly. "You all have to be properly debriefed."

"Forget it," she said. "This is it, Dr. Carmichael. I can tell you everything there is to tell, and I'm not going to give you any names. You can go home to Washington but I live here. There really isn't anything to hunt around for. Even the chimps are incinerated—

nothing left but a handful of bone-ash and a few teeth. You'll have to start all over again, somewhere else. I'm sorry, but that's the way it is."

"Okay," he said. "Tell me what you know—about the project, and about the progress Abel and Franklin had made."

She nodded. "I know about the jigsaw hypothesis," she told him. "I may have been only a makeshift technician, but I know the general outline of it. The entire solar system is made out of cosmic wreckage, right? All the heavy elements are debris from a series of burned out stars. Abel thought that DNA was debris too—that there's tons of the stuff out in deep space. He thinks that the evolution of life on earth has been—what did he call it?—a re-collection. He thought that there were pathways already laid down for natural selection, because what it's really doing isn't building from scratch but re-building something that already existed once before, millions of years ago and thousands of light years away. Like putting together a jigsaw of DNA. He thought that we might already have the next evolutionary phase mapped out within us, didn't he? He was trying to make those chimps evolve—and his ultimate aim was to try it out on people too, to turn us into supermen overnight. Except that I saw those chimps every day, Dr. Carmichael, and whatever it was they'd become, it wasn't human. They changed, all right—but not into people."

Carmichael was relieved to discover that she knew as much as she did. It made her evidence that much more useful, that much more reliable. Even so, she didn't quite have the idea in her grasp. "There's no reason to think the chimps would evolve into people," he told her, scrupulously. "Chimps aren't one step below us on the evolutionary ladder. They're our cousins—practically our brothers."

He saw a gleam in her eye, as if she thought she'd scored a point. Perhaps it had all been guesswork, and she hadn't really been sure until he confirmed it. Perhaps she didn't know as much as she was pretending to...or perhaps there was some subtler game that she was playing.

"They weren't any smarter, so far as I could tell," she said. "Just weird-looking. The lost their hair, and their skin got thicker, like a rhino's. They grew taller and more thick-set, less playful. They got to like me while I was feeding them, before the experiment started, but afterwards they got confused—maybe depressed. Abel kept trying to talk to them in Ameslan, but he didn't seem to be getting anywhere. It's possible that the whole thing was a dead end—if it wasn't, he never got the chance to make the breakthrough."

"But his methods seem to have worked out," Carmichael said. "That's a breakthrough in itself—and the chimps were still alive after they changed, still functioning. You probably don't realize how significant that is."

Lucy Vollman shrugged. "Maybe if you tell me more," she said, "I'll be able to tell you more. Maybe it'll trigger something."

"Okay," said Carmichael, a little warily. Was there any reason not to tell her? It wasn't war work, after all. He drew a deep breath, and continued. "As you say, Abel's so-called jigsaw hypothesis proposes that the DNA making up earth's biosphere is recapitulating, at least in broad terms, an evolutionary sequence that's already taken place elsewhere in the universe, perhaps on millions of other planets orbiting other secondary stars—that is to say, in other systems formed out of the debris of ancient supernovas. If that were true, the process of evolution on any particular world would be analogous to the way in which plant species invade virgin ground, each species in its turn transforming the environment so as to create niches for other invaders, until you eventually reach a mature climax community.

"Some people who find Abel's hypothesis attractive think that human beings are the climax community of Earthly evolution; others think there might be one or two pieces of the genetic jigsaw left to slot into place. Abel's own opinion seemed to be that we might be so very close that it might only be a matter of switching on and off a few genes already *in situ* among the quiet DNA of our chromosomes. The differences between humans and chimpanzees aren't the result of our possession of genes that they don't have, but simply the result of the differential switching of genes we already hold in common.

"We already have techniques for controlling the expression of genes in primitive organisms; Abel was trying to use those techniques on chimps. He wasn't trying to humanize them, at least in any narrow sense of the term—he was trying to explore the potential they had for becoming...well, maybe in some respects *better than human*. He was trying to find in them the potential that might be in us for completing the jigsaw, and bringing the story of our evolution to its allotted climax. Not that it was ever likely to be that simple, of course; we have hundreds of thousands of years of cultural evolution behind us, and that's what has shaped our minds. Our genes only shape our bodies and our brains. It was all just a hypothesis anyway—maybe just a fairy story.

"I've read the reports Abel managed to file, but they don't track the metamorphosis as far as you've observed it; they only mention a few relatively subtle physiological changes—interesting enough, in their way, but not proof of anything. Unless you can give us a strong reason for continuing the work, the project will probably die, at least until the war ends. Millions of people have died since Abel and Franklin moved out here, and we have more urgent priorities now. Even so, I have a responsibility to salvage whatever information I can, and file it away in the hope that when better times come it will give someone else a flying start. Can you add anything further?"

The gleam in her eye was still there. He hadn't told her anything she didn't know—was she pleased that he didn't know any more?

"I don't think so," she said. "I wish I could tell you something really exciting, but I can't. As I say, the chimps seemed to me to get stupider rather than smarter, and if the next step in evolution involves growing that kind of skin, I'll be happy to stay primitive. I can't say anything about physiology and biochemistry, but lookswise those apes were going backwards, not forwards. I guess we'll have to win the war before we try again—if we can win the war. If things go on the way they are, we might just lose everything. Wouldn't it be ironic if, just as we were about to put the last piece in place, the whole damn jigsaw fell apart? If we go, the rats and the cockroaches will have to start over with the business of re-collection."

"It's only a hypothesis," he reminded her. "Nothing you've told me counts as proof. What I'd like you to do, if you will, is write down everything you remember: everything, including things that may not seem important. Will you do that for me?"

She shrugged. "Sure," she said. "I'll get on it tomorrow." After a pause, she added: "I'd better go now."

He nodded, although she really didn't need his permission. She put her wet coat back on, and left the way she'd come in. He didn't stir until she was out of sight. Then he leaned over, picked up the phone from beside the bed, and dialed the lobby. Andrews was waiting by the phone, and picked up immediately.

"It's okay," said the sergeant. "We saw her climbing the fire-escape. I sent Kravitz after her. He'll find out where she's going, and report back."

"Good," said Carmichael. "Smart work, sergeant." It was, of course, no more than he'd expected, given the sergeant's entirely understandable determination to be watchful at all times.

* * * * * * *

Carmichael spent the following morning at the police lab, going through the reports of the investigating officers and the results of the various tests carried out by Burke's team. The atmosphere around him was uncomfortably frosty; everyone was scrupulously polite but they all felt that his presence implied some criticism of their work. He was checking up on them, and they didn't like it. He didn't like it any better than they did. It was a relief when Andrews came to tell him, shortly after noon, that Kravitz had returned.

"What the hell took him so long?" Carmichael asked.

"It was a long trip," said Andrews, laconically. "She doesn't live in town—at least, she didn't stay in town when she left your room. She went way up into the hills. Walked all the way. Kravitz nearly lost her two or three times, but he had the IR image-enhancer from his rifle-sight, and he picked her up again."

"She went back to the burned-out labs?" said Carmichael, puzzled.

Andrews shook his head. "Cabin in the woods," he said. "Miles away from the installation—from anything. Guy let her in—Kravitz didn't get a good look at him then, but he hung around for a while. There are at least three people living there. One of them's tall, slim, fair-haired, fiftyish.

Daniel Franklin was—or had been—tall, slim, fair-haired and fiftyish. So were lots of other people. Carmichael rooted out photographs of Franklin and Abel, and told Andrews to show them to Kravitz while he finished up what he was doing.

He dropped in to see Burke before he left. "You did a good job," he said. "Nobody could have done better. You didn't conclude anything you weren't able to confirm—I appreciate your caution."

Burke didn't look pleased; he thought it was just a line, to soften him up. "That's okay," he said, dully.

"About the apes," said Carmichael. "Your best guess is that there were two corpses, not three?"

"No," said Burke. "There wasn't enough to justify any kind of educated guess. The remains we found were consonant with the theory that there might have been two non-human primate corpses, which might or might not have been chimpanzees. There were anomalies. I wouldn't swear to the number or the nature, or that the bones we found came from anything that was alive before the fire. I didn't draw any conclusions at all, Dr. Carmichael."

Carmichael nodded. "You're right," he said. "Too little to go on. Whoever set the fire knew how to make it burn exceedingly hot—anything left could be disinformation. Even Abel's teeth."

"You think Abel set the fire?" said Burke.

"I don't think anything," Carmichael told him. "I'm on your side—no conclusions, no guesses, no theories. It's the only scrupulous way."

When Carmichael got back to the hotel, he found Kravitz in a similar sort of mood, unable to confirm or deny that the man he'd seen at the cabin was Daniel Franklin. "Didn't see him clear enough," he said, unapologetically. "Could have been the guy in the photograph, but it's a lousy picture and I was looking through a night-sight.

"Do we tell the local cops about the cabin?" Andrews wanted to know.

"Not yet," said Carmichael. "The people up there sent the girl down to talk to me, avoiding the locals. It's just possible that the local cops are the ones they're hiding out from—and if so, it might be as well to know why before we blow their cover. We'll go up there, just the three of us. Discreetly."

"When?"

"After dark. And we'll walk—leave the jeep where it is."

Andrews didn't look entirely happy about it, but he shrugged his shoulders. Kravitz, on the other hand, looked distinctly miserable. He'd already had one sleepless night. Carmichael relented. "Okay," he said to the soldier. "Just the sergeant and me—provided that you can show us on the map exactly where this cabin is, and how to approach it."

Carmichael went back to his routine duties then, going through the motions with studied efficiency, even though he no longer expected to turn up anything of interest. It was difficult to concentrate, but he was a patient man and he was careful to give no sign of the fact that he thought he might have found a better route to the answers for which he had been sent to hunt.

He and Andrews set off from the hotel at ten; there was no evidence that they were being watched or followed. The sergeant wore a sidearm but didn't take a rifle; they both carried night-sights. They didn't have to go far, in terms of miles on the map, but it was mostly uphill and the rain had left the ground muddy and slick. Carmichael found that he wasn't quite as fit as he'd assumed. By the time they got to the trail out to the cabin he was breathing hard.

The advice Kravitz had given them was easy enough to follow, and they were able to approach the cabin stealthily enough—but when they got there it seemed that there had been no need for stealth. It was dark and seemed to be deserted.

The door was locked but the enterprising Sergeant Andrews got them inside without using brute force. The interior was neat enough, but there was no real sign of contemporary occupation. It was like hundreds of other homes which had been abandoned in the course of the plague war. It had not been looted—but it was no treasure-house by anyone's standards, and looters could afford to pick and choose nowadays.

"Looks like they moved on," said Andrews, when they'd briefly checked all the rooms." He wasn't the kind to insult his own man by wondering aloud whether Kravitz had somehow contrived to direct them to the wrong cabin.

"Leaving precious little behind," Carmichael agreed, sourly. He knew that he'd have to search more thoroughly, looking for something that might prove that Franklin had been here, and he knew only too well how tedious such a search might be.

"I'll take a look around outside," said the sergeant, probably feeling that it would be diplomatic to give Carmichael some space.

Carmichael nodded. He didn't have equipment with him for dusting for fingerprints—he'd have to come back later, or ask Burke for help, if it became necessary to look for evidence of that kind. All he could do for now was look for clues of a grosser kind. He started in the larger bedroom, checking the floor and the drawers. After ten or twelve minutes without any result he moved into the kitchen.

When he heard the front door open he assumed it was the sergeant coming back. He didn't bother to call out as footsteps approached the kitchen door, which stood ajar, but simply waited until the door swung inwards before looking round.

The man who stood in the doorway looking in wasn't Andrews—it was a man in a police uniform. The expression on the cop's face was difficult to read, but Carmichael realized that his presence in the cabin might take some explaining.

"It's okay, officer," he began, as the policeman drew his gun. "It's...."

While he was speaking his arms spread out reflexively, palms wide open to indicate his harmlessness. He realized with less than a second to spare that the gesture wasn't having the desired effect—that the cop was going to shoot. He had no time to scream for help; he barely managed a plaintive "Hey!"

The gun made a curious spitting sound, which somehow seemed to make the whole experience rather surreal. Carmichael looked down at his chest, and saw some kind of dart sticking out of his shirt, with a red stain slowing spreading around it.

It's not a bullet, he thought, wonderingly. The incident seemed almost as bizarre as it would have if the gun had released a flag with BANG! printed on it.

Carmichael felt his amazement turning into giddiness, and swayed to his left. He had time to catch himself up once, before swaying the other way. This time he couldn't help himself, and he was conscious of slowly falling over, crumpling at the knees. No sooner had he hit the ground than consciousness fizzled out.

* * * * * * *

He awoke in a dimly lit room, slumped in an old armchair; it didn't take his eyes long to adjust to the light, but there was a roaring pain in is head and he squinted in the hope that it might help the pain to go away.

"Drink this," said a voice, pressing a cup to his lips. It was cool water. After he'd taken a sip the cup was placed in his hand, and when his eyes were fully open he saw two white tablets in an open hand. He looked up at the man who was offering them to him.

"Aspirin," said the blond-haired man. "Good for your head and your heart."

Carmichael took the tablets and swallowed them, washing them down with the water. As he did so he looked around. There were two other people present: a second man, and Lucy Vollman. The second man was dark, shorter than the other. Like Kravitz before him, Carmichael couldn't tell from the photographs he'd seen whether the blond man was Franklin—nor could he tell for certain whether the dark man was Abel. He handed the cup back to the man who stood over him.

"There was no need for this," he said, glancing at he girl. "If you wanted to talk, all you had to do was call."

The girl looked down at her fingernails, but she didn't seem particularly guilty about having lied to him.

"We have to be discreet, Dr. Carmichael," said the dark man. "These are troubled times."

Carmichael noted that the speaker seemed to have a full set of teeth, which—if they weren't false—were in remarkably good condition for a man of his age. "Dr. Abel?" he said, experimentally.

The dark man grinned. "Am I still recognizable?" he asked. "Or is it just that the situation is conducive to jumping to that conclusion?"

"You'll have to do better than a little superficial somatic engineering," said Carmichael. "Your genetic fingerprint will still be the same. Even if you've done a little creative gene-switching, you can't change that. And the fact that you left a few teeth behind for the investigators to find does rather imply that you weren't an innocent victim of the attack on your labs. Perhaps, like Dr. Franklin here, you should have been content to leave nothing. What did you do with Sergeant Andrews, by the way?"

"Sergeant Andrews hasn't suffered any permanent damage," said Abel, enigmatically. "Even his ego will recover, in time—and he doesn't ever have to know what happened to him, if you decide not to tell him."

"Actually," added the blond man, "the fire might have been a little too effective. We had hoped to convince the investigators that we were both dead. But we didn't start the fire, and we didn't leave any of our own people to burn. We simply had enough advance warning to make a few preparations. The human bones were those of plague-victims—easy enough to acquire around these parts. We hoped that everybody would believe that the bombers had succeeded—including, of course, the bombers."

"So your story is that the bombers really were local anti-biotech fanatics?" said Carmichael, skeptically. *They might be telling the truth*, he thought, *but so far, they've conducted themselves in a manner which isn't conducive to belief in anything they may say.*

"They certainly had some local help," said Abel, "but we're really not certain who was behind it, in the ultimate analysis. Do you know, Dr. Carmichael?"

It took Carmichael a second or two to realize that if the question was sincere, Abel must mean to imply that the order might have come out of Washington. Even that was possible, and it was possible that the people who'd sent Carmichael out to check up hadn't leveled with him as to the reason for which he was being sent. It seemed unlikely, though—and one had to be careful not to get too paranoid, lest one's left hand should become absurdly suspicious of one's right.

"I'm not here to find out who did it," Carmichael said, cautiously. "I'm here to retrieve any information that can be retrieved regarding your research. For purely scientific reasons, we'd like to know how far advanced your project was." He glanced around at the room. It was nondescript—just another abandoned room in another abandoned house. The ragged curtains weren't completely closed, but beyond them the cloudy night was still pitch-dark.

"No doubt you were anxious," Abel purred, "because I was a little dilatory about filing my reports." Carmichael glanced at Lucy Vollman again, knowing that she must have repeated everything he'd told her, more or less word for word. This time she met his gaze steadily enough.

"Aren't we all?" he said.

"Not when they concern the war situation," said Abel. "Not that ours did, of course, except in a rather oblique fashion. Ordinarily, I might have been quite prepared to publish what we found, at the risk of meeting some hostility and ridicule, but in the present situation, I must confess, I was inclined to hesitate—and was nearly lost. You must forgive us our extreme caution, Dr. Carmichael. These days, anyone who isn't paranoid is certifiably insane—isn't that what they say, back in Washington?"

And everywhere else, thought Carmichael. His head was clearer now, and not so painful. He sat up straighter. "Did you bring me here to give me an explanation in confidence?" he asked, sharply. "Or are you intending to hold me for ransom?"

"We brought you here because we need help," said Abel. "Help that you can provide."

"You want me to help you get another lab?" said Carmichael, disbelievingly.

Abel didn't answer the question. Instead, he said: "Lucy says you know all about the jigsaw hypothesis—chapter and verse. You were well briefed; you've read the reports I did file."

Carmichael nodded.

"In that case, perhaps I can ask you to speculate a little. Suppose, for the sake of argument, that the jigsaw hypothesis were true—that evolution on earth has been rebuilding a close approximation of a path that DNA has already followed elsewhere in the universe, and that human beings are just one significant step short of becoming the true kingpins of the climax community. Tell me, Dr. Carmichael: what characteristics would you hope or expect to find in the ultimate hominids, our destined—and I do mean *destined*—successors?"

It was on the tip of Carmichael's tongue to say that the ability to grow a new set of teeth probably wouldn't come amiss, but Abel was clearly serious and there was no point in frivolity. "I don't know," he said, warily. "The kind of things, I suppose, that genetic engineers would like to build into us: longer life, better powers of self-repair, better immunity to diseases."

"And better looks," said Franklin, dryly. "Don't forget better looks. That's why we've made such rapid progress in superficial somatic engineering, after all. Cosmetic engineering was the big business, until the war broke out."

"Everybody wants to look like a demigod," said Carmichael. "So what?"

"That's right," said Abel. "We've always made our gods in our own image, according to our own ideals of beauty, and we've tended to assume that any progress built into the evolutionary scheme would take us in the direction of our ambitions and aspirations. If we really were to take control of the evolutionary process, that's what we'd do. But even if what we're doing, here on Earth, is mere recollection and recapitulation, what it's recollecting and recapitulating is the result of natural selection, where hopes and ambitions have to submit to the rigors of the struggle for existence. Would you like to see our successor species, Dr. Carmichael? Would you like a glimpse into the future which awaits us, if we can't and don't take control of our own destiny?"

So it wasn't all lies, Carmichael thought. *She didn't lie about the chimps.* Aloud, he said: "Okay." He began to rise from his seat, but Franklin gestured him back again, before striding to the door.

"You can come in now, Mike," he said to someone waiting in another room. Apparently, the person he had spoken to was the man who had shot Carmichael back at the cabin, and he still had the gun in his hand, but he was no longer wearing the uniform to which he wasn't entitled. Presumably, he had the gun ready because the creature which was with him wasn't under any kind of restraint, although it seemed quite docile. It walked through the doorway ahead of the man, and stood there staring at Carmichael. There was curiosity in its stare, but no apparent malice.

It was about five feet tall, and it walked more like a man than a chimp, although it was round-shouldered. Its features were vaguely Neanderthal—pronounced brow-ridges, bull-neck, peculiar teeth visible behind rubbery lips—but it wasn't hairy. Its hide bore only the faintest resemblance to that of a rhinoceros, but Carmichael could see how Lucy Vollman had made the connection. It was thick and tough and dark. It reminded Carmichael of the rubbery suits film-makers used to contrive in Hollywood when they had to dress some luckless actor up as a thing from outer space. The hide was smooth, and there were no external genitalia—only a shapeless fold of skin that just might have been a natural codpiece. The creature's eyes were very narrow, with slit-like pupils, more like a snake's than a man's.

Carmichael and the creature looked one another over for a couple of minutes. Carmichael had to make an effort to still the beating of his anxious heart as that alien gaze bore into him, but the thing that had been a chimpanzee seemed quite relaxed.

"That's okay, Steve," said Abel, with a faint grin teasing the corners of his mouth. "You and Mike can go get dinner now." He looked at the man, so that Carmichael would know that he'd been tempted to a false conclusion when Franklin spoke earlier. "Mike" was actually the ex-chimp. As if to underline the point, Abel added: "Thanks, Mike."

The monster nodded, gestured with his oddly slender fingers, and turned away.

"He's really pretty bright," said Abel. "We thought at first his IQ had gone down rather steeply, but he was probably just confused. Of course, we don't know what he might have been capable of if he'd been born into a community in the normal way. He's in an anomalous situation; it's as if he were a feral child brought into a society of aliens. Maybe his kind have the potential for speech;

maybe he'd be very smart indeed if he hadn't spent most of his life as an ape—we're not sure. Without trying the process on a human being—preferably a baby—we'll probably never be sure, but somehow I don't think we'd get the experiment past an ethics committee, do you?"

"Just because you switched on a few quiet genes," said Carmichael, "it doesn't mean that you're looking into the evolutionary future. That's just a freak—it doesn't mean a thing."

"That's possible," Abel conceded. "But on the other hand, Mike and his two companions have certain traits which are not uninteresting, and far from useless. That skin of his may not be what the fashionable demigod is wearing this year, but it's one hell of a tough tegument. It's a far more effective shield than your skin or mine against all kinds of harmful radiation, against barrier-transmissible poisons of the class which includes many of our best nerve-gases, and against most common-or-garden instruments of assault. It won't necessarily stop a bullet, but it's at least as good as the army-issue flak-jacket we took from your sergeant. As for things that do manage to get inside, one way or another...well, Mike has a truly ferocious immune system. It doesn't come without cost, of course. As far as we can tell, Mike ages at least fifty per cent faster than his old self—more than twice as fast as you or me—but while he lives, nothing much can touch him. As I say, we don't really have much of an idea of how intelligent he'd be if his mental potential were fully-developed, but physically he's a lot closer to the superman than we are."

It wasn't difficult to see where the argument was pointing. "The sort of people our remote ancestors might become," Carmichael said, "given the right—or do I mean wrong?—environment. Civilization cracked apart, no longer able to sustain its productive base; soils poisoned by pollution; cancer and mutation rates running riot; decimation by plagues; war, war and more war. Back to the Stone Age, with all kinds of added hazards."

"That's an oversimplified view," Franklin put in, scrupulously. "It's possible, you see, that *Homo sapiens* was always an aberration, a dead-end sideline. Maybe the sequence was supposed to run from the guys that were common ancestors to the chimps and ourselves more-or-less straight through to people like Mike. That's one of the neat little touches to the jigsaw hypothesis, you see—the picture you'd get on any particular world would build up differently, depending on the order in which you put the pieces in, and what opportunities there might be for fitting in pieces that don't actually belong in the final, completed puzzle. In the end, though, there's only one satisfactory answer."

"I don't believe that," Carmichael said. "We're on the threshold of taking charge of our own evolution. If we can only become competent genetic engineers, we can become anything we want to be—including demigods."

"The question," said Abel, "is whether we can cross that threshold before the whole edifice comes crashing down around our ears. My guess is—and I speak from bitter experience—that we'll be burned out before that happens. My guess is that the whole fucking human species is burned out like a dying leper, and that the plague war is just one of a host of symptoms...just one of a whole horde of marauding horsemen of the Apocalypse."

"So what do you want to do?" said Carmichael, spitting the words out. "Found a colony of Mikes, hiding out from the end of the human world?"

"Pretty much," said Abel, unperturbed by the sarcasm or the contempt. "I know it takes a lot more swallowing than an aspirin, but all you need is time to think it through. All those backwoods morons who think they're survivalists have got the right idea but they don't have the means or the will to carry it through. Given the opportunity, we just might. All we need is a couple of million dollars' worth of equipment, and some real security. When I say *real*, I mean *really* real. We want in on the bunker culture, Dr. Carmichael. We want a place in the best bolt-hole in the world. And we want you to persuade the people who matter that we deserve it."

"What makes you think that I'm in a position to do that?" Carmichael asked.

"We don't know that you are. In a way, we hope that maybe you aren't—quite. But you and I once worked in the same building, although you probably don't remember. I was an outsider, working along very speculative lines, but you were an insider even then. You might be lower down the totem pole than you'd like to be, but you know where the ladders are, and how to buy yourself a slot. I'm doing you a favor, and I think you'll be able to see that, when you've had time to think it over. I'm well up to date with my reports, by the way—it's just that I haven't been filing them for a while. Nothing really worthwhile was lost in the fire. As I say, we had advance warning that something of the sort was due to happen. But if your people want to see my reports, they'll have to come up with an offer we like. I'm not just going to hand them over—and the best bits are staying right in here." He tapped the side of his head.

Carmichael thought of saying something along the lines of "You're crazy," but he knew there'd be no point. Everybody was crazy these days. These were crazy times.

Instead, he sighed, and said: "Where exactly are we?"

Abel smiled. In fact, they all smiled. They thought they were winning. Perhaps they were. He'd have to think it over—but in the meantime, there was nothing to do but play along. He'd got what he came for, and he'd have time enough to think about what he actually wanted, and what it might be possible to get.

* * * * * * *

When Carmichael got back to the hotel he found Sergeant Andrews in a very bad mood indeed: bitterly embarrassed and even more bitterly angry. There were two local detectives with him, but his own men were wisely keeping their distance.

"They jumped me," he explained, awkwardly. "There were three, I think—but they shouldn't have got close."

"Survivalists," said one of the detectives. "They're trained, and they probably knew the territory. They were probably after the gun, the night-sight and the flak-jacket rather than the money-belt, but of course they took that too. You'd think you'd be safe inside the city limits, but nobody is—not these days."

"It wasn't complacency," said Andrews, through gritted teeth. Carmichael didn't doubt it for a second. "Doc, I need to take a couple of the men..."

"That wouldn't be a good idea," said the second detective, swiftly.

"No," said Carmichael, softly. "I don't suppose it would. I'm sorry, sergeant, but I can't do that. We don't have the time. We have to go back east tonight."

Andrews looked genuinely surprised. "But, Doc...!" he protested. He was too good a man to say anything about the girl in front of the local men, but his eyes spoke volumes.

"I'm sorry," said Carmichael again, before turning to the policemen. He assured them that there would be no problems, that it would all be left to them—but that Washington would be in touch again if they couldn't make progress. He added a remark to the effect that if civil society really had broken down to the extent that was apparent, it might be time to consider martial law. The cops' grins turned into half-scowls as they caught his implications, but they left without starting any argument.

"It's okay," said Carmichael, raising his hand as Andrews opened his mouth to speak. "I'll take the responsibility for the equipment and the money. I'll tell them that I recklessly ordered you into a dangerous situation against your better judgment, and that I told Kravitz to stay behind, leaving you without back-up. Your superiors will be only too happy to blame it all on some dumb-ass scientist."

Andrews looked uncertain.

"I don't want the local police asking too many questions," Carmichael added. "I don't want them to know that I found the girl. She's scared of the people who burned down the labs, and she doesn't trust the police. I can't blame her."

"You saw her?" Andrews seemed surprised.

Carmichael nodded. "I looked for you outside," he lied, "but I couldn't hang around to search the bushes. Anyhow, I saw the girl again—and two of the other people who worked up at the labs. I got a full enough account of what was going on—full enough to make up into a passable report. That's all we're here for. Given the way things are around here, I think I'd just as soon get back home as soon as humanly possible—know what I mean?"

The sergeant still looked uncertain, but he nodded again. He felt the back of his head, where he'd been hit. "Feels like enemy territory," he said. "I mean, this is the USA, right? We're supposed to be on the same fucking side."

"Things are falling apart," said Carmichael, sympathetically. "So many people are dying that the infrastructure is collapsing. It's not safe to drink the tap-water, and they can't even keep the TV networks going. Sheer cultural inertia is all that's holding the Union in place, and the barbarians are at the gates, looting, pillaging and burning. The survivalists might be just a little bit ahead of their time; the war of all against all could be just about to begin."

"Hell," said the sergeant, "I never thought to see it in my lifetime—leastways, not unless the nukes started falling. But I guess that could happen too, if we ever figure out just who it is that's attacking us. If things go on the way they are...."

"They'll have contingency plans back in Washington," said Carmichael. "In Washington, they have contingency plans for everything. That's the place to be, when the going gets tough. Let's get back there as soon as we can, hey?"

The sergeant stood up. "I'll tell the men," he said. "How soon do you want to hit the road?"

"I've got nothing else to do here," Carmichael told him. "I can be ready in an hour."

"You look a little rough," Andrews observed. "Not as rough as me, but rough enough."

"Lack of sleep," said Carmichael, although his fingers came up reflexively to touch the wound on his chest where the anesthetic dart had hit him. "It doesn't matter. I'll catch a little in the back of the lorry, once we're on the road to the airfield."

The sergeant nodded, and went to the door. As he opened it, he looked back. "This really was a wild goose chase, wasn't it?" he asked. "There was nothing here to find—we were just sent out here to put up some kind of show."

"I'm afraid so," said Carmichael, consciencelessly.

"Sending us out on a job like that was a pretty stupid thing to do, wasn't it?" said Andrews, bitterly. "Considering how ugly things are getting."

"I guess it was," Carmichael agreed, off-handedly. He was already beginning to think hard about how much of what Abel and Franklin had told him was really believable, and how much was likely to be lies—and what, in any case, he ought to tell his superiors back east. It was a difficult puzzle, with many facets, but he had every confidence that in due course he would find the best way to fit the pieces together. The best way, that is, for him. As he had just told Sergeant Andrews, the war of all against all was just about to begin.

Andrews was still looking at him, and Carmichael realized that he had let the last sentence dangle, as though he were about to follow it up with some profound remark.

"But things could get uglier yet," he added, regretfully. "A whole lot uglier."

INHERIT THE EARTH

Damon Hart never found it easy to get three boxes of groceries from the trunk of his car to his thirteenth floor apartment; it was a logistical problem with no easy solution, given that both his parking-slot and his apartment door were so far from the elevator. Some day, he supposed, he would have to invest in a collapsible electric cart, but such a purchase still seemed like another step in the long march to conformism—perhaps the one which would finally seal his fate.

By the time he opened the apartment door he felt distinctly ragged. He could have done without the carving-knife that slammed into the door-jamb ten centimeters away from his ducking head and stuck there, quivering.

"You bastard!" Diana said, rushing forward to meet him.

It didn't take much imagination to figure out what had offended her so deeply. He should have tidied the work away, concealing it behind some gnomic password.

"It's not a final cut," he told her, setting the first box down and raising his arms with the palms flat in a placatory gesture. "It's just a first draft. It won't be you in the finished product—it won't be anything like you."

"That's bullshit," she said, her voice still taut with pent-up anger. "First draft, final cut—I don't give a damn about that. It's sick, Damon."

He knew that it might add further fuel to her wrath but he deliberately turned his back on her and went back into the corridor to fetch the second box of groceries. *This is it*, he thought, as he picked it up. *This is really it*. In a ideal world there ought to be a more civilized way of breaking up, but theirs had always been a combative affair, whose every stress and strain became manifest in explosive anger. In the beginning, that had added excitement, but things had now reached the stage when all the storm and stress was a burden he could do without.

My fighting days are over, Damon thought. *I can't do it any more.*

Once the last box was inside and the door was safely closed behind him, he felt that he was ready to face her. Her tremulous rage was already dissolving into tears and she was digging her fingernails into her palms so deeply that they were drawing blood. With Diana, violence always shifted abruptly into a masochistic phase; real pain was sometimes the only thing that could block out the kinds of distress with which her internal technology was not equipped to deal.

"You don't want me at all," she complained. "You don't want any living partner. You only want my virtual shadow. You want a programmed slave, so you can be absolute master of your paltry sensations. That's all you've ever wanted."

"It's a commission," he retorted, bluntly. "It's not a composition for art's sake, or for my own gratification. It's not even technically challenging. It's just a piece of work. I'm using your body-template because it's the only one I have that's pre-programmed to a suitable level of complexity. Once I've got the basic script in place I'll modify it out of all recognition—every feature, every contour, every dimension. I'm doing it this way because it's the easiest way to do it. All I'm doing is constructing a pattern of appearances; it's not real."

"You don't have any sensitivity at all, do you?" she came back. "To you, the templates you made of me are just something to be used in petty pornography. They're just something convenient. It wouldn't make any difference what kind of tape you were making, would it? You've got my image worked out to a higher degree of digital definition than any other, so you put it to whatever use you can: sextapes, horrorshows...anything. It really doesn't matter to you whether you're making training tapes for surgeons or masturbation-aids for freaks, does it?"

As she spoke she struck out with her fists at various parts of his imaging system: the console, the screens and—most frequently—the dark helmet within whose inner surface a clever programmer could inscribe an infinite range of imaginary worlds.

"I can't turn down commissions," said Damon, as patiently as he could. "I need connections in the marketplace and I need to be given problems to solve. Yes, I want to do it all: sextapes and training tapes, abstracts and dramas, games and repros and stupid ads. I want to be master of it all, because if I don't have all the skills, anything I devise for myself will be tied down by the limits of my own idiosyncrasy."

"And templating me was just another exercise. Building me into your machinery was just a way to practice."

"It's not you, Di," he said, wishing that he could make her understand. "It's not your shadow, certainly not your soul. It's just an appearance. When I use it in my work I'm not using *you*."

"Oh no?" she said. "When you stick your head into that black hole and put that plastic suit on, you leave this world behind and you enter another. When you're there—and you sure as hell aren't *here* very often—the only contact you have with me is with my appearance, and what you do to that appearance is what you do to me. When you put my image through the kind of motions you're building into that sleazy fantasy you're designing it's me you're doing it to, and no one else."

"When it's finished, he said, doggedly, "it won't look or feel anything like you. Would you rather I paid a copyright fee to reproduce some stock character? Would you rather I sealed myself away for hours on end with a hired model?"

"I'd rather you spent some time with me," she told him. "I'd rather you lived in the actual world instead of devoting yourself entirely to substitutes. I never realized that giving up fighting meant giving up life."

"You had no right to put the hood on," Damon told her, coldly. "I can't work properly if I feel that you're looking over my shoulder all the time. That's worse than knowing that I might have to duck when I come through the door because you could be waiting for me with a deadly weapon."

"It's only a kitchen-knife. At the worst it would have put your eye out."

"I can't afford to take a week off work while I grow a new eye—and I don't find experiences like that amusing or instructive."

"You were always too much of a coward to be a first-rate fighter," she told him, trying hard to be scornful. "You switched to the technical side of the business because you couldn't take the cuts."

Damon had never been a reckless fighter, all flamboyance and devil-may-care; he had always fought to win with the minimum of effort and the minimum of personal injury. Because most of his opponents hadn't cared much about skill, or art, or even sensible self-preservation, he had won four out of five of his fights. He didn't consider that to be evidence of stupidity or stubbornness—and he'd switched to tape-doctoring because it was more challenging and more interesting than carving people up.

"If you want the sound and fury of the streets," he said, tiredly, "you know where they are."

"You don't need me any more, do you?" she complained. "All you ever wanted of me is in that template. As long as you have my appearance programmed into your private world you can do anything you like with me, without ever having to worry whether I'll step out of line. You'd rather have a virtual image than a real person, wouldn't you? You wouldn't even take that helmet off to eat and drink if you didn't have to. If you had any idea how much you've changed since...."

It was probably truer than she thought, but he didn't see any need to be ashamed of it. The whole point about the world inside a VR hood, backed up by the full panoply of suit-induced tactile sensation, was that it was better than the real world: brighter, cleaner and more controllable. Earth wasn't Hell any more, thanks to the New Reproductive System and the wonders of internal technology, but it wasn't Heaven either. Heaven was something a man could only hope to find on the other side of experience, in the virtuous world of virtual imagery.

The brutal truth of the matter, Damon thought, was that everything of Diana Caisson that he actually needed really was programmed into her template. The absence from his life of her changeable, complaining, untrustworthy, knife-throwing self wouldn't leave a yawning gap.

"You're right," he told her. "I've changed. So have you. That's okay. We're authentically young; we're supposed to change. We're supposed to become different people, to try out all the personalities of which we're capable. The time for constancy is a long way ahead of us yet." He wondered, as he said it, whether it was true. Was this really just a phase in an evolutionary process, or was it a permanent capitulation? Was he taking a rest from the kind of hyped-up sensation-seeking existence he'd led while running with Madoc Tamlin's gang, or was he turning into one of the meek whose alleged destiny was to inherit the earth?

"I want the template back," Diana said, sharply. "When I go, I'm taking my virtual shadow with me."

"You can't do that," Damon retorted, knowing that he had to put on the appearance of a fight before he eventually gave in, lest it be too obvious that all he had to do was remold her simulacrum from the modified echoes which he had built into half a dozen different commercial tapes of various kinds. While he only required her image, he could always get her back, no matter how comprehensively and how ostentatiously he purged his systems of her likeness.

"I'm doing it," she told him, firmly. "You're going to have to start that slimy sideshow from scratch, whether you pay for a readymade template or pay for some whore who'll let you build a new one on your own."

"If I'd known," he said, thinking to it was probably safe now to be calculatedly provocative, "I wouldn't have had to struggle upstairs with three boxes of groceries."

From there, it was only a few more steps to a renewal of the armed struggle, but he kept the knife out of it, and his aim—as always—was to win with the minimum of fuss. He made her work hard to dispel her bad feeling in pain and physical stress, but she got there in the end, without having to scream too much abuse.

Afterwards, he helped her pack.

There wasn't that much to collect. It only filled four boxes—and because there were two of them to do the work, they didn't pose that much of a logistical problem.

When he got back to the apartment, the cops were waiting for him.

* * * * * * *

Damon knew that it couldn't be a trivial matter if the cops had taken the trouble to call in person. Even the cops conducted their interviews by video, unless they had some special reason for appearing in person.

"Whatever it is," Damon was quick to say, as a smartcard identifying the senior man as Inspector Hiru Yamanaka of Interpol was held out for his inspection, "I'm not involved. I don't run with the gang any more and I don't have any idea what they're up to. These days, I only go out to fetch the groceries."

The men from Interpol preceded Damon into the apartment, ignoring the stream of protestations. Yamanaka showed not a flicker of interest as his heavy-lidded gaze took in the knife stuck into the door-jamb but his unnamed sidekick took silent but ostentatious offence at the untidy state of the room.

As soon as the door was shut Yamanaka said: "What do you know about the Eliminators, Mr. Hart?"

"I was never that kind of crazy," Damon told him, affrontedly. "I was a serious street-fighter, not a hobbyist assassin."

"No one's accusing you of anything," the second cop said, in the unreliably casual way cops had.

Damon knew no more about the Eliminators than anyone else—perhaps less, given that he was no passionate follower of the kinds of newstape that followed their activities with avid fascination. He was not entirely unsympathetic to those who thought it direly unjust that longevity, the pain-control, immunity to disease and resistance to injury were simply commodities to be bought off the nanotech shelf, possessed in the fullest measure only by the rich, but he certainly wasn't sufficiently hung up about it to become a terrorist crusader.

The Eliminators were on the lunatic fringe of the many disparate and disorganized communities of interest fostered by the Web; they were devoted to the business of giving earnest consideration to the question of who might actually deserve to live forever. Some of their so-called Operators were in the habit of naming those whom they considered "unworthy of eternity", via messages dispatched to netboards from illicit temporary linkpoints, usually accompanied by downloadable packages of "evidence" which put the case for elimination. The first few freelance executions had unleashed a tide of media alarm—which had, of course, served to glamorize the whole enterprise and conjure into being a veritable legion of amateur assassins. Being named by a well-known operator was not yet a guarantee that a person would be attacked and perhaps killed, but it was something that had to be taken seriously.

It didn't take much imagination to understand that Interpol must be keen to nail a few guilty parties and impose some severe punitive sanctions, *pour encourager les autres*, but Damon couldn't begin to figure out why their suspicions might have turned in his direction.

"May I?" Yamanaka asked. His neatly-manicured finger was pointing to the windowscreen.

"Be my guest," Damon said, sourly.

Yamanaka's fingers did a brief dance on the windowscreen's keyboard. The resting display gave way to a pattern of words etched blue on black:

CONRAD HELIER IS NAMED AN ENEMY OF MANKIND
CONRAD HELIER IS NOT DEAD
FIND AND IDENTIFY THE MAN WHO WAS CONRAD HELIER
PROOFS WILL FOLLOW
OPERATOR 101

Damon felt a sinking sensation in his belly. He knew that he ought to have been able to regard the message with complete indifference, but the simple fact was that he couldn't.

"What has that to do with me?" he asked, combatively.

"According to our records," Yamanaka said, smoothly, "you didn't adopt your present name until ten years ago, when you were in your teens. Before that, you were known as Damon Helier. You're Conrad Helier's natural son."

"So what? He died twenty years before I was born, no matter what that crazy says. We're about to begin the twenty-third century—it doesn't matter any more who anybody's natural father was."

"To most people," Yamanaka agreed, "it's a complete irrelevance—but not to you, Mr. Hart. You were given your father's surname. Your four foster-parents were all close colleagues of your father. Your father even left money in trust for you, which you inherited a couple of years after changing your name. I know that you've never touched the money, and that you don't see your foster-parents any longer, and that you've done your utmost to distance yourself from the destiny which your father apparently planned out for you—but that doesn't signify irrelevance, Mr. Hart. It suggests that you took a strong dislike to your father and everything he stood for."

"So you think I might do something like this? I'm not that stupid, and I'm certainly not that crazy. Who put you on to me? Who pointed the finger at me? Was it Karol Kachellek?"

"No one identified you as a possible suspect," the Interpol man said. "We're checking everyone who might have some kind of grudge against Conrad Helier—or Conrad Helier's memory. We know that Operator 101 always transmits his denunciations from the L.A. area, and you've been living hereabouts throughout the time he's been active."

"I told you—I'm not that kind of lunatic, and I try never to think about Conrad Helier and the plans he had for me. I'm my own man, and I have my own life to lead. Why are you so interested in a message that's so patently false? You can't possibly believe that Conrad Helier is still alive—or that he was an enemy of mankind, whatever that's supposed to mean."

"We're interested because it's a new departure," Yamanaka said, evenly. "No operator, including 101 has ever used the phrase *enemy of mankind* before. Nor has any ever appealed to kindred spirits to do anything other than kill someone. It might be a hoax, of course—one of the nastiest aspects of the Eliminators' game is that anyone can play. This code-number's been used eight times but that doesn't necessarily mean that all the messages came from one source. We became even more interested when we began checking it out. You're not the only person connected with Conrad Helier living hereabouts."

"One of my foster-parents, Silas Arnett, lives south of San Francisco," Damon admitted. "I haven't seen him in years. We don't communicate at all."

"He seems to have disappeared from his home," Yamanaka said. "We don't know how long he's been missing, but we fear foul play. Another of your father's contemporaries, Surinder Nahal—the only person who might conceivably qualify as an enemy of his, according to our files—has an address in San Diego, but he's proving equally difficult to trace."

"I don't know Nahal at all," Damon said, truthfully.

"Karol Kachellek also claimed that he hadn't seen Arnett for many years," Yamanaka added. "Eveline Hywood said the same. It seems that your surviving foster-parents fell out with one another as well as with you."

"Maybe—Silas's decision to retire must have seemed to Karol and Eveline to be a failure of vocation almost as worthy of censure as my own: yet another betrayal of Conrad Helier's sacred cause."

Yamanaka nodded, as if he understood—but Damon knew that he almost certainly didn't. It was difficult to guess Yamanaka's true age, because a man of his standing would have the kind of internal technology that was supposedly capable of sustaining eternal youth, but Damon judged that he was no centenarian. To the policeman, as to Damon, Conrad Helier's career would be the stuff of history. At school he must have been told that the artificial wombs that Conrad Helier had perfected, and the techniques that had allowed such wombs to produce legions of healthy infants while the plague of sterility spread like wildfire across the globe, were the salvation of the species—but that didn't mean that he understood the reverence in which Conrad Helier had been held by his co-workers.

"Do you have any idea why anyone would want to blacken your father's name fifty years after his death, Mr. Hart?" Yamanaka asked, with a blandness that was patently false.

"I was encouraged in every possible way to see my father as the greatest hero and saint the world ever had," Damon said. "I couldn't follow in his footsteps, and I didn't want to, but that doesn't mean that I disapprove of where they led. Whoever posted this notice is sick."

"There were several witnesses to the death of Conrad Helier," the Interpol man said, matter-of-factly. "The doctor who was in attendance and the embalmer who prepared the body for the funeral both confirm that they carried out DNA-checks on the corpse, and that the gene-map matched Conrad Helier's records. If the man whose body was cremated on 27 January 2147 wasn't Conrad Helier, then the gene-map on file in the Central Directory must have been substituted." He paused briefly, and then said: "You don't look at all like your father. Is that deliberate, or is it simply that you resemble your mother?"

"I've never gone in for cosmetic reconstruction," Damon told him, warily. "I have no idea what my mother looked like; I don't even know her name. I understand that her ova were stripped and frozen at the peak of the Crisis, when people were afraid that the world's entire stock might be wiped out by the plague. There's no surviving record of her name. At that time, according to my co-parents, nobody was overly particular about where healthy ova came from; they just wanted to get as many as they could in the bank. They were stripping them from anyone more than five years old, so it's possible that my mother was a mere infant."

"It's possible, then, that your natural mother is still alive," Yamanaka observed.

"If she is," Damon pointed out, "she can't possibly know that her ovum was inseminated by Conrad Helier's sperm and that I was the result."

"I suppose Eveline Hywood and Mary Hallam must both have been infected before their wombs could be stripped," Yamanaka said, disregarding the taboos that would presumably continue to inhibit free conversation regarding the legacy of the plague until the last survivors of that era had retired from public life. "Or was it that Conrad Helier was reluctant to select one of your co-parents as a natural mother in case it affected the partnership?"

"I don't think any of this is relevant to the matters you're investigating," Damon said, "and I think the investigation itself is a waste of time."

"I don't know what might be relevant and what might not," Yamanaka said, unapologetically. "The message supposedly deposited by Operator 101 might be pure froth, and there might be nothing sinister in the disappearance of Arnett and Nahal—but this could represent the beginning of a new and nastier phase of Eliminator activity. They already attract far too much media attention, and this story is only one step short of becoming a headline scandal. I have to investigate it at least as assiduously as the dozens of newsmen who must have been commissioned to start digging, and I need to stay at least one step ahead of them. I'm sorry to have troubled you, but I judged it to be necessary, if only to inform you of what had happened."

He's delicately implying that I might be in danger, Damon thought. *If he's right, and Silas's disappearance has something to do with the message, this really might be the beginning of something nasty. Even if it's only a newstape hatchet job....*

"I'll ask around," he said, carefully. "If I discover anything that might help you, I'll be sure to let you know."

"Thank you, Mr. Hart," the man from Interpol said, offering no clue as to exactly what he understood by Damon's promise to ask around. "I'm grateful for your co-operation."

* * * * * * *

"It's too tight," the boy complained. "I can't move properly."

"No it's not," said Madoc Tamlin, with careful patience, as he knelt to complete the synaptic links in the *reta mirabile* that covered the fighter's body like a bright spider web. "It's no tighter than pants and a shirt. You can move quite freely."

The boy's fearful eyes looked over Tamlin's shoulder, lighting on Damon's face. Damon saw the sudden blaze of belated recognition. "Hey," the boy said, "you're Damon Hart! I got a dozen of your fight tapes. You going to be doing the tape for this? That's great! My name's Lenny Garon."

Damon didn't bother to interrupt the flow to tell the kid that he hadn't come to watch the fight or that he hadn't—as yet—been contracted to doctor the tapes. He understood how scared the youngster must be. The fight was only for show, but that didn't mean the kid wasn't going to get hurt; in fact, if Damon judged the situation rightly, it was a guarantee that he was going to get cut up pretty badly. It would be the first time the little sucker had gone up against a skilled knifeman, and he must know that he was out of his depth. There was a certain irony in the fact that the only way someone like Lenny Garon could make enough money to equip himself with tissue-repair technology was to sustain the injuries that the relevant nanotech was geared to undo, but Damon no longer found anything to savor in that kind of irony.

Tamlin stood up, already issuing stern instructions as to where the combatants shouldn't stab one another. He didn't want the recording apparatus damaged. "The only way you can make real money for this kind of work," he told the kid, "is to get used to the kit and to make damn sure it doesn't get damaged. Given that your chances of long-term survival are directly proportional to your upgrade prospects, you'd better get this right, because it's the only break you're likely to get. Savvy?"

Garon nodded dumbly. Tamlin was a major player in the Underworld games these days, and the boy respected his opinion. "I can do it," he said, uneasily. "I got all the feints and jumps. It'll be okay."

"We don't want feints and jumps" Tamlin countered, with a contemptuous sneer intended to wind the boy up. "We want purpose and skill and desperation. Just because we're making a VR tape...explain it to him, Damon."

"We aren't making a simple recording that will give a floater the illusion that he's going through your moves, Lenny," Damon said, off-handedly. "We're making a template. It's raw material, which will have to be carefully refined, but it has to have a sense of urgency about it—an edge. Play-acting doesn't do the job. It reeks of fake. I know it's difficult, but if you want to be good at this, you have to go all the way...and as Madoc says, you have to look after the wiring. No record at all is even worse than a bad one."

The kid nodded respectfully. Damon still had a reputation on the streets; his tapes made sure of that.

"Just remember," Tamlin said, as he pushed the boy forward, "it's a small price to pay for taking one more step towards immortality." Street parlance always spoke of immortality rather than emortality—which, strictly speaking, was all that even the very best internal technology could hope to provide.

Damon watched two fighters square up. Their kit was more than a little cumbersome, but very few artificial organics were as delicate as the real thing and you couldn't get template-precision with thinner webs. Then he looked away, at the ruined buildings to either side of the street. This whole district was ex-urban wilderness, emptied by the Crash and never re-colonized or reclaimed. Nobody lived here; it was just a vast playground for the gangs. Damon wondered what it must have been like in the bad old days of the Crisis, crowded out with the unemployable and the insupportable: one of countless concentration-city powder-kegs waiting for an inflammatory spark that had never come. He couldn't imagine it. Even the very old, who had lived through the Crisis and the Crash, had mostly lost their memories of it.

The fight itself was boring, although the other watchers—whose sole *raison d'être* was to whip the combatants into a frenzy—weighed in with the customary verve and fury. Amazingly, the kid managed to stick Brady in the gut while the experienced fighter was playing cat and mouse with him—which made Brady understandably furious. It was immediately clear that he wasn't going to settle for some token belly-wound as a reprisal; he wanted copious bloodshed. That would be more than okay by Tamlin, so long as the cuts didn't do too much damage to the recorders. Young Lenny would be all the more enthusiastic to volunteer for something really heavy in order to pay for the nanotech that would make him as good as new and keep him that way.

Tamlin noticed Damon's reluctance to join in the loud exhortations of the crowd. "Don't get all stiff on me, Damon," he said.

"You may be in the Big World now, but you're too young to get *rigor mortis*. You pissed about splitting with Diana?" Tamlin hadn't said so, but Damon presumed that Diana had gone straight to him after the split. Tamlin surely wouldn't take her back on a full-time basis, but he'd be ready to lend her a shoulder to cry on, for a week or two.

"Interpol came to call," Damon told him, abstractedly. "They were asking about Eliminators."

"Eliminators! You don't have any truck with them, do you?"

"Of course not. It's just a connection from the distant past—something I thought I'd left way behind me."

Madoc Tamlin was the only person Damon knew who would be able to take the correct inference from those words; even Diana Caisson didn't know that Damon Hart had once been Damon Helier. Damon saw the flicker of interest ignite in his friend's eye, but Tamlin knew better than to say too much out loud. All he said was: "Oh?"

"You know some light-footed Webwalkers, don't you?" Damon said. "Do you know anyone who could do Interpol-type work better than Interpol can?"

"They all say they can," Tamlin replied, cautiously. "It's a key item of the creed that all the best cracksmen are outlaws. The really good ones get all their commissions from the corps, though—they're just undercover suits with expensive tastes. I don't know anyone who could outsmart Interpol on the cheap. Nobody does."

"If this thing turns out to be serious," Damon said, stressing the *if*, "I'd be willing to lay out serious credit to pursue it."

"Do you have that kind of money?" Tamlin asked, warily. "You haven't been making it on what I pay you."

"I've got some put away," Damon said, feeling no compulsion to specify where it had come from. He fetched a smartcard out of his pocket and held it out. "It's already authorized for cash withdrawals," he said. "It's all above board. You can draw ten thou with no questions asked. If you need more, call me—but it had better be worth paying for."

"What am I looking for?" Tamlin asked, mildly.

"Some local pervert calling himself Operator 101 has posted a notice about Conrad Helier, claiming that he's still alive and that he's guilty of some as-yet-unspecified crime. One of my foster-parents, Silas Arnett, has gone missing from home near San Francisco."

"I thought you didn't like your foster-parents," Tamlin said, keeping one eye on the fight. Lenny Garon was in real trouble now. The crowd were baying for blood, and getting it. Damon kept his own eyes firmly on Tamlin's face.

"We had a disagreement," Damon said, dismissively. "They only did what they thought was right, and Silas tried a lot harder than Karol or Eveline to figure out what that might involve. After I left, he dropped out too. I owe him."

"It's Helier you're really interested in, isn't it?" Tamlin asked, running his fingers speculatively back and forth along the edge of the smartcard. "This Arnett guy is a side-issue. You want to know if your natural father really is alive."

"If he were," Damon admitted, "I'd like to know. But what I really want to know is whether he really was an enemy of mankind." He said it lightly, to imply that he was joking, but he knew that Tamlin would wonder whether this was one of the many true words that were rumored to be spoken in jest. He wasn't entirely sure himself.

"How are things otherwise?" Tamlin asked, finally putting the smartcard away. "Honest toil living up to your expectations?"

"I'm taking a break," Damon told him. "A brief excursion to Hawaii."

"Vacation?"

"Independent line of enquiry. Karol Kachellek is there, working out of Molokai. He probably won't tell me anything, even if he has some idea what's going on, but if I go in person I might at least unsettle him a bit."

Tamlin shrugged.

"You might also ask around about someone named Surinder Nahal," Damon added. "He was a bioengineer contemporary with my co-parents. He seems to have been a rival of sorts—maybe the closest thing to an enemy they had. He's disappeared too."

Tamlin nodded, and then turned back abruptly to the fight as a roar went up from the watchers. Brady had rammed his advantage home, and poor Lenny Garon was on the ground, screaming.

Damon knew that it would all be feeding into the template: the reflexes and convulsions of pain, shock and horror, all ready-digitized, ripe for manipulation and refinement, for teasing into proper shape. By the time he—or someone like him—had finished with the tapes there'd be nothing of the kid left at all; there'd only be the actions and the reactions, dissected out and purified as a marketable commodity. It was all in rank bad taste, of course, but it was a living for all concerned. It was his own living, based in talents that were entirely and exclusively his own, using nothing that Conrad Helier had left to him in his will or his genes.

Damon wanted very badly to be his own man. Taking money from the legacy to bankroll Tamlin's investigations in the Underworld wasn't a betrayal of that end; it was an utterly impersonal matter. It seemed wholly appropriate that Conrad Helier's money should be used to find out what was going on—always assuming, of course, that something worthy of investigation was going on.

Tamlin had moved forward to help the stricken street-fighter—not because he was overly concerned for the boy's health but because he wanted to make certain that the equipment was in good order.

"Give my regards to Diana," Damon said, as he turned away. "Tell her I'm sorry, but that it'll all work out for the best."

He couldn't tell whether Tamlin had heard him or not.

* * * * * * *

Damon stood on the quay in Kaunakakai's main harbor and watched the oceanographic research vessel *Kite* sail smoothly towards the shore. The wind was light and her engines were silent but she was making good headway. Her sleek sails were patterned in red and yellow, shining brightly in the subtropical sun. Karol Kachellek didn't come up to the deck until the boat was coming about, carefully shedding speed so that she could drift to the quay. He didn't wave a greeting and he kept Damon waiting while he supervised the unloading of a series of cases, which presumably held samples or specimens. Two battered trucks with low-grade organic engines had already limped down to the quayside to pick up whatever the boat had brought in.

Eventually, Kachellek came over to Damon and offered his hand to be shaken. Kachellek had always been distant; Silas Arnett had been the real foster-father of the group to whose care Damon had been delivered in accordance with his father's will, just as poor dead Mary had been the real foster-mother.

"This isn't a good time for visiting, Damon," Kachellek said. "We're very busy." At least he had the grace to look slightly guilty as he said it. He raised a hand to smooth back his unruly blond hair. "Let's walk along the shore," he went on. "It'll be a while before the mud samples are ready for examination, and there won't be any more coming in today. Things might be easier in three or four weeks, if I can get more staff, but until then...."

"You're very busy," Damon finished for him. "You're not worried, then, by the news?"

"I haven't the time to worry about Eliminators and other assorted madmen. I can't see why Interpol's personnel are so excited about a stupid message cooked up by some sick mind. It should be ignored, treated with the contempt it deserves. Even to acknowledge its existence is an encouragement to further idiocies of the same kind."

They soon passed beyond the limits of the harbor, and headed westwards towards the outskirts of the port. Mauna Loa was visible in the distance, looming over the precipitous landscape, but the town itself was uncomfortably reminiscent of the parts of L.A. where Damon had spent the greater part of his adolescence.

Molokai had been one of the many bolt-holes whose inhabitants had tried to impose quarantine in the early 2100s, but the plague had arrived here as surely as it had arrived everywhere else. Artificial wombs had been imported on the scale that the islanders could afford, but that wasn't large, and the population of the whole chain had been dwindling ever since. The internal technologies that guaranteed longevity to those who could afford them would have to become even cheaper before that trend went into reverse, unless there was a sudden influx of immigrants. In the meantime, the part of the port that remained alive and active was surrounded by a ragged halo of concrete wastelands.

There was little to see on the landward side but the lingering legacy of human profligacy, so Damon looked out to sea while they walked. The ocean gave the impression of having always been the way it was: huge, blue and serene. Where its waves lapped the shore they created their own dominion, shaping the sandy strand and discarding their own litter of wrack and rot-misshapen wood. Lanai was visible on the horizon, on the far side of the Kaiohi Channel

"You and Silas were friends for a long time," Damon remarked. "Aren't you concerned about his disappearance?"

The blond man shrugged. "We were colleagues, not friends," he said. "When we ceased to be colleagues we ceased to mean anything to one another. People live for a long time nowadays, Damon, and it's no longer the case that people you know for fifty or a hundred years have to play a major role in the unfolding narrative of your life. Whatever has happened to Silas, it doesn't have anything to do with me."

It was too stark and too brutal to be entirely convincing. Kachellek had never been entirely at ease with Damon, so it was difficult to judge whether he was any more unsettled than usual, but there was something about his dismissiveness that seemed dishonest.

I must be careful of seeing what I want to see, Damon thought. *I must be careful of wanting to find a juicy mystery, or evidence that my paternal idol had feet of tawdry clay.* "Do you feel the same way about Eveline?" he said, aloud. "Was she just someone you worked with for a while, before you went your separate ways?"

"I'm still working with her," Kachellek replied.

"But she's off-planet, in L-5."

"Modern communications make it easy enough to work in close association with people anywhere in the solar system. We're involved with the same problems, constantly exchanging information. In spite of the hundreds of thousands of miles that lie between us, Eveline and I are close in a way that Silas and I never were. We're in harmony, dedicated to a common cause." It was a horribly pompous speech; Kachellek was by no means a subtle man.

"A common cause that I deserted," Damon said, bitterly, "in spite of all the grand plans that Conrad Helier had for me. You're not him, Karol. You could have seen my point of view, if you'd wanted to. You and I could have built a relationship of our own."

"Fostering you was a job your father asked me to do," Kachellek retorted, bluntly. "I'd have continued doing it, if there had been anything more I could do—but what you wanted was to get away, to abandon everything your father tried to pass on to you, to run wild. You moved away from us, Damon, and changed your name; you declared yourself irrelevant to our concerns. I don't owe you anything."

Damon didn't want to become sidetracked into discussions of his irresponsible adolescence, or his not-entirely-respectable present. "Why should someone accuse Conrad Helier of being an enemy of mankind?" he asked.

"He's dead, Damon," Kachellek said, softly. "Nobody can hurt him, whatever lies they make up."

"They can hurt you and Eveline. They might have hurt Silas already. Surely that's reason enough to be interested, if not afraid? Whatever they're planning to say about my father will reflect on you too—unless you think he's just another colleague you happened to work with once upon a time, whose acquaintance has now become irrelevant."

"Conrad can never be irrelevant to me," Kachellek said, rising obediently to the bait but not showing the slightest sign of bad temper. "He isn't able to work on the problem that faces us just now, but he's present in spirit in every logical move I make, every hypothesis I frame, every experiment I design. He made me what I am, just as he made the whole world what it is. You and I are both his heirs, and we'll never be anything else, however hard we try to avoid the consequences of that fact." He obviously had no intention of giving Damon an easy ride.

Kachellek paused before a rocky outcrop that was blocking their path, and knelt down as if to duck any further questions. He scanned the tide-line, which ran along the wave-smoothed rock a few inches above the ground. The weed that was clinging there was slowly drying out in the sun, but the incoming tide would return before it was desiccated; in the meantime, the limp tresses provided shelter for tiny crabs and whelks. Where the weed was interrupted, sea-anemones nestled in crevices like blobs of jelly. The bare rock above the tide-line was speckled with colored patches of lichen and tarry streaks which might have been anything. Kachellek took a pen-knife from his pocket and scraped some of the tarry stuff from the rock into the palm of his hand, inspecting it carefully. Eventually, he tipped it into Damon's hand and said: "That's more important than all this nonsense about Eliminators."

"What is it?" Damon asked.

"We don't have a name for the species yet—nor the genus, nor even the family. It's a colonial organism reminiscent in some ways of a slime-mold. It has a motile form that wanders around by means of protoplasmic streaming, but the colonies can also set rock-hard. Its genetic transactions are inordinately complicated and so far very mysterious—but that's not surprising, given that it's not DNA-based. Its methods of protein-synthesis are quite different from ours, based in a radically different genetic system and genetic code."

Damon had given up genetics, and had carefully set aside much of what his co-parents had tried assiduously to teach him, but he understood the implications of what Kachellek was saying. "Is it new," he asked, "or just something we managed to overlook during the last couple of centuries?"

"We can't be absolutely certain," said Kachellek, scrupulously. "But we're reasonably certain that it wasn't here before. It's a recent arrival in the littoral zone, and so far it hasn't been reported anywhere outside these islands."

"So where did it come from?"

"We don't know yet. "The obvious contenders are up, down...." He seemed to be on the point of adding a third alternative, but didn't; instead he went on: "I'm looking downwards; Eveline's investigating the other direction.

Damon knew that he was expected to rise to the challenge and follow the line of argument. The *Kite* had been dredging mud from the ocean bed, and Eveline Hywood was in the L-5 space-colony. "You think it might have evolved on the sea-bed," Damon said. "Maybe it's been there all along, since DNA itself evolved, or maybe not. Perhaps it started off in one of those bizarre enclaves that surround the black smokers where the tectonic plates are pulling apart and has only just begun expanding its territory, the way DNA did a couple of billion years ago. On the other hand, maybe it drifted into local space from elsewhere in the universe, in the form of Arrhenius spores...again, maybe a long, long time ago or maybe the day before yesterday. How different from DNA is its replicatory system?"

"We're still trying to confirm a formula," Kachellek told him. "We've slipped into the habit of calling it para-DNA but it's a lousy name because it implies that it's a near chemical relative, and it's not. It coils like DNA—it's definitely a double helix of some kind—but its subunits are quite different. It seems highly unlikely that the two have a common ancestor, even at the most fundamental level of chemical evolution. It's a separate creation. That's not so surprising—whenever and wherever life first evolved there would surely have been several competing systems, and there's no reason to suppose that one of them would be superior in all conceivable environments. The hot vents down in the ocean depths are a different world—life down there is chemosynthetic and thermosynthetic rather than photosynthetic. Maybe there was always room down there for more than one chemistry of life. Perhaps there are other kinds still down there—that's what I'm trying to find out. In the meantime, Eveline's looking at dust samples brought in by probes from the outer solar system. The system is full of junk, and it's not beyond the bounds of possibility that life has evolved in the outer regions, or that spores of some kind could have drifted in from other systems. We don't know—yet."

"You don't think this stuff poses any kind of threat, do you?" said Damon, intrigued in spite of himself. "It's not likely to start displacing DNA organisms?"

"Until we know more about it," Kachellek said, sternly, "it's difficult to know how far it might spread. It's not likely to pose any kind of threat to human beings, given the kind of nanotech defenses we can now muster, but that's not why it's important. Its mere existence expands the horizons of the imagination by an order of magnitude. What are a few crazy slanderers, even if they're capable of inspiring a few crazy gunmen, compared with this?"

"If it is natural," said Damon, "it could be the basis of a whole new spectrum of nanomachines."

"It's not obvious that there'd be huge potential in that," Kachellek countered. "So far, this stuff hasn't done much in the way of duplicating the achievements of life as we know it, let alone doing things that life as we know it has never accomplished. It might be inefficient, capable of performing a limited repertoire of self-replicating tricks with no particular skill; if so, it would be technologically useless, however interesting it might be in terms of pure science. We're not looking to make another fortune, Damon—when I say this is important, I don't mean commercially."

"I never doubted it for a moment," Damon said, dryly—and turned abruptly to look at the man who was rapidly coming up behind them. For a moment, it crossed his mind that this might be an Eliminator foot-soldier, mad and homicidal—but he was an islander, and Kachellek obviously knew him well.

"You'd better come quick, Karol," the man said. "There's something you need to see. You too, Mr. Hart."

* * * * * * *

The package had been dumped into the Web in hypercondensed form like any other item of mail, but once it had been downloaded and unraveled it played for a couple of hours of real time. It had been heavily edited, so the claim with which it was prefaced—that nothing had been altered or falsified—couldn't be taken seriously.

The material was addressed TO ALL LOVERS OF JUSTICE and it was titled ABSOLUTE PROOF THAT CONRAD HELIER IS AN ENEMY OF MANKIND. It came—or purported to come—from the mysterious Operator 101. Kachellek and Damon watched in anxious silence as it played back.

The first few minutes of film showed a man bound to a huge, throne-like chair. His wrists and ankles were pinned by two pairs of plastic sheaths, each three centimeters broad, which clasped him more tightly if he struggled against them. He was in a sitting position, his head held upright by a device which neatly enfolded his skull. His eyes were covered but his nose, mouth and chin were visible. His pelvic region was also enclosed.

"This man," a voice-over announced, "is Silas Arnett, an intimate friend and close colleague of Conrad Helier. He has been imprisoned thus for seventy-two hours, during which time almost all of his protective nanomachinery has been carefully flushed from his body. He is no longer protected against injury, nor can he control pain."

Damon glanced sideways at Kachellek, whose face had set like stone. He didn't doubt that this was, indeed, Silas Arnett; nor did he doubt that Arnett had been stripped of the apparatus that normally protected him against injury, aging and the effects of torture.

But if they intend to force some kind of confession out of him, Damon thought, *everyone will know that it's quite useless. Take away a man's ability to control pain and he can be made to say anything at all. What kind of "absolute proof" is that?*

The image abruptly shifted to display a virtual courtroom. It was a highly impressionistic image—a cartoon rather than a serious attempt at videosynthesis. The accused man who stood in the dock was a caricature, but Damon had no difficulty in recognizing him as Silas Arnett. The twelve jurors who were positioned to his left were mere sketches, and the person whose position was to the right—presumably the prosecutor—had little more in the way of features than they did. The black-robed judge who faced Arnett didn't look any more real, but in his case it wasn't for lack of detail; his face had been carefully drawn, and Damon's expert eye judged that it had been carefully designed for convincing animation.

"Please state your name for the record," said the judge. His voice was deep and obviously synthetic.

"I'll do no such thing," said the figure in the dock. Damon recognized Silas Arnett's voice, but in the circumstances he couldn't be sure that the words hadn't been synthesized by a program that had analyzed the voice.

"Let the name Silas Arnett be entered in the record," said the judge. "I am obliged to point out, Dr. Arnett, that there really is a record. Every moment of this trial will be preserved for posterity.

Any and all of your testimony may be broadcast, so you should conduct yourself as though the whole world were watching. Given the nature of the charges which will be brought against you, that might well be the case."

"I didn't think you people bothered with interrogations," Arnett said. It seemed to Damon that he was injecting as much contempt into his voice as he could. "I thought you operated strictly on a *sentence first, verdict afterwards* basis."

"It sometimes happens," said the judge, "that we are certain of one man's guilt, but do not know the extent to which his collaborators and accomplices were involved in his crime. In such cases we are obliged to undertake further enquiries."

"Like the witch-hunters of old," said Arnett, grimly. "I suppose it would make it easier to select future victims if the people you select out for murder were forced to denounce others before they die. Any testimony you get by such means is worse than worthless; this is a farce, and you know it."

"We know the truth," said the judge, flatly. "Your role is merely to confirm what we know."

"Fuck you," Arnett said, with feeling. It was a direly old-fashioned curse: something out of Conrad Helier's Ark, which should not have any force in the modern world. The significance of the word, and the act it described, had changed considerably since the old world died; the word had lost the warrant of obscenity it had once possessed.

"The charges laid against you, Silas Arnett, are these," said the machine-made voice, while the judge's virtual lips moved in perfect sync. "First, that in the years 2070-2080 you did conspire with others, including Conrad Helier, Mary Hallam, Eveline Hywood and Karol Kachellek, to cause actual bodily harm to between thirteen billion and fifteen billion individuals by disabling their reproductive organs. Second, that you did collaborate with others, including those named, in the design, manufacture and distribution of the various virus species known as meiotic disrupters and chiasmalytics. How do you plead to these charges?"

Damon was astonished by his own reaction, which was more extreme than he could have anticipated. He was seized by an actual physical shock that left him trembling. He turned to look at Karol Kachellek, but the blond man wouldn't meet his eye. Kachellek seemed remarkably unperturbed, considering that he had just been accused of manufacturing and spreading the great plague of sterility whose dire effects he and his collaborators had so gloriously subverted.

"If you had any real evidence," Arnett said, while the face of his simulacrum took on a strangely haunted look, "you'd have brought these charges in a real court of law. The simple fact that I'm here demonstrates the absurdity and falseness of any charges you might bring."

"You've had a hundred and twenty years to surrender yourself to judgment by another court," said the judge, his voice acidly mechanical. "This court is the one that has found the means to bring you to trial; it is the one that will judge you now. You will be given every opportunity to enter a defense, should you so wish and to justify your criminal actions as best you can, before sentence is passed upon you."

"I refuse to pander to your delusions. I've nothing to say."

"Our investigations will be scrupulous nevertheless," the judge said. "They must be, given that the charges, if true, require sentence of death to be passed upon you."

"You have no right to do that!"

"On the contrary: we hold that what society bestows upon the individual, through the medium of technology, society has every right to withdraw from those who betray their obligations to others. This court intends to investigate the charges laid against you as fully as it can, and when they are proven it will invite any and all interested parties to pursue those who ought to be standing beside you in the dock. None will escape, no matter what lengths they may have gone to in the hope of evading judgment. There is no station of civilization sufficiently distant, no deception sufficiently secure, to place a suspect beyond our reach."

What's that supposed to mean? Damon wondered. *Do they really believe that Conrad Helier is alive, and that he faked his death in order to avoid punishment for what he'd done?* "Karol..."

he said aloud—but Kachellek raised a hand to instruct him to be quiet.

"The people you've named are entirely innocent of any crime," Arnett said, anxiously. "You're insane if you think otherwise."

Damon tried to judge from the timbre of the voice the extent to which Arnett's pain-control system had been dismantled, so as to force him to suffer physical symptoms of distress to which he had long been unaccustomed. If there were indeed a reality behind this cartoonish charade Arnett's body must by now be an empire at war, and he must be feeling all the violence of the conflict. The tireless molecular agents that benignly regulated the cellular commerce of his emortality were dying beneath the onslaught of custom-designed assassins: Eliminators in miniature, which would exterminate them all, given time, and leave their detritus to be flushed out by his kidneys. Even if Arnett had not yet been subjected to actual torture he must feel the returning grip of his own mortality, and the deadly cargo of terror which came with it.

The picture dissolved, and was replaced by an image of Conrad Helier, which Damon immediately recognized as a famous section of archive footage.

"We must regard this new plague not as a catastrophe, but as a challenge," Helier stated, in ringing tones. "It is not, as the Gaian Mystics would have us believe, the vengeance of Mother Earth upon her rapists and polluters, and no matter how fast and how far it spreads it cannot and will not destroy the species. Its advent requires a monumental effort from us, but we are capable of making that effort. We have, at least in embryo, technologies that are capable of rendering us immune to aging, and we are rapidly developing technologies that will allow us to achieve in the laboratory what fewer and fewer women are capable of doing outside it: conceive and bear children. Within twenty or thirty years we will have what our ancestors never achieved: democratic control over human fertility, based in a New Reproductive System. We have been forced to this pass by evil circumstance, but let us not undervalue it; it is a crucial step forward in the evolution of the species, without which the gifts of longevity and perpetual youth might have proved a double-edged sword...."

The speech faded out. It was easy enough for Damon to figure out why the clip had been inserted. Recontextualized by the accusations that the anonymous judge had brought against Arnett, it implied that Conrad Helier had thought the great plague a good thing: an opportunity rather than a curse; a significant step on the road to salvation.

Damon had no alternative but to ask himself the questions demanded by the mysterious Operator. Had Conrad Helier been capable of designing the agents of the plague as well as the instruments which had blunted its effects? If capable, might he have been of a mind to do it? The answer to the first question, he was certain in his own mind, was *yes*. He was not nearly as certain that the answer to the second question was *no*—but he had never known his biological father; all he had ever known was the oppressive force of his father's plans for him, his father's hopes for him. He had rebelled against those, but his rebellion couldn't possibly commit him to believing this—and he did know the other people named by the judge. Karol was distant and diffident, Eveline haughty and high-handed, but Silas and Mary had been...everything he required of them. Surely it was unimaginable that they could have done what they now stood accused of doing?

The image cut back to the courtroom, but the moment Damon heard Silas Arnett speak he knew that time had elapsed. A section had been cut from the tape.

"What do you want from me?" Arnett hissed, in a voice full of pain and exhaustion. "What the fuck do you want?"

It was not the virtual judge who replied this time, although there was no reason to think that the second synthesized voice issued from a different source. "I want to know whose idea it was to launch the great plague," said the figure to Arnett's right. "I want to know where I can find incontrovertible evidence of the extent of the conspiracy. I want to know the names of everyone who was involved. I want to know where Conrad Helier is now, and what name he's using."

"Conrad's dead. I saw him die!" Arnett's voice was almost hysterical, but he seemed to making Herculean efforts to control himself.

"No you didn't," said the accusing voice, without the slightest hint of doubt. "Someone switched the DNA-samples so as to fake the identification. Was that you, Dr. Arnett?"

There was no immediate reply. The tape was interrupted again; there was no attempt to conceal the cut. Damon could imagine the sound of Arnett's screams easily enough; only the day before he had listened to poor Lenny Garon recording a tape that it would probably be his privilege to edit and doctor and convert into a peculiar kind of art....

"It was my idea," Silas Arnett said, in a hollow, grating voice. "Mine. I did it. The others never knew. I used them, but they never knew."

"They all knew," said the inquisitor firmly.

"No they didn't," Arnett insisted. "They trusted me, absolutely. They never knew. They still don't—the ones who are still alive, that is. I did it on my own. I designed the plague and set it free, so that Conrad could do what he had to do. He never knew that it wasn't natural. He died not knowing. He really did die not knowing."

"That's very noble of you," said the other, in a voice dripping with irony. "But it's not true, is it?"

"Yes," said Arnett.

This time, the editor didn't bother to cut out the sound of screaming. Damon shivered, even though he knew that he and everyone else who had managed to download the tape before Interpol deleted it was being manipulated. This was melodrama, not news—but how many people, in today's world, could tell the difference? How many people would be able to say: *It's just some VR pornotape. It doesn't mean a thing.*

Suddenly, Diana Caisson's reaction to the discovery that Damon was using her template as a base for the sextape he had been commissioned to make didn't seem quite so unreasonable.

The interrogator spoke again. "The truth, Dr. Arnett, is that at least five persons held a secret conference in May 2072, when Conrad Helier laid out his plan for the so-called salvation of the world. The first experiments with the perfected viruses were carried out in the winter of 2075-76, using rats, mice and human tissue-cultures. When one of his collaborators—was it you, Dr. Arnett?—asked

Conrad Helier whether he had the right to play God his reply was 'The position's vacant. If we don't, who will?' That's the truth, Dr. Arnett, isn't it? Isn't that exactly what he said?"

Arnett's reply to that was unexpected. "Who are you?" he asked, his pain seemingly mingled with suspicion. "I know you, don't I? If I saw your real face, I'd recognize it, wouldn't I?"

The answer was equally surprising. "Of course you would," the other said, with transparently false gentleness. "And I know you, Silas Arnett. I know more about you than you can possibly imagine. That's why you can't hide what you know."

At this point, without any warning, the picture cut out. It was replaced by a text display, which said:

CONRAD HELIER IS AN ENEMY OF MANKIND
FIND AND IDENTIFY CONRAD HELIER
MORE PROOFS WILL FOLLOW
OPERATOR 101

Damon stared numbly at the words, which glowed as if they had been written in fire.

* * * * * * *

When Damon called Madoc Tamlin, Diana answered; mercifully, it was only a voice link, so neither of them had to look the other in the eye.

"It's Damon," he said, brusquely. "I need to get a message to Madoc. Tell him I really need that package we discussed. He has to get on to it right away. I've authorized him to draw twice as much cash. I'll fly back tonight or tomorrow, and I really need to know whatever he can dig up as soon as humanly possible. Have you got all that?"

"Of course I've got it," she snapped back. "Do you think I'm stupid or something?"

Damon had to suppress an impulse to react in kind. Instead, he said: "Sorry, Di—I'm a bit wound up. Just ask Madoc to do what he can, and tell him he has extra resources if he needs them, okay?"

By the time he signed off Karol Kachellek had put the other phone receiver down. Damon didn't know who he had been talking to. "I'm sorry, Damon," Kachellek said, bluntly. "You were right—this is far worse than I thought. It couldn't have come at a worse time."

"What's it all about, Karol?" Damon asked, quietly. "You do know, don't you?"

"I wish I did. You mustn't worry, Damon. It will all be sorted out. I don't know who's doing this, or why, but...." As the blond man trailed off, Damon stared at him intently, wondering whether the red flush about his brow and neck was significant of anger, anxiety, embarrassment or some combination of all three.

Kachellek reddened more deeply under his gaze. "It's all lies, Damon," he said, awkwardly. "You can't possibly believe any of that stuff. They forced Silas to say what he did. We can't even be sure that it really was his voice. It could all have been synthesized."

"It doesn't much matter whether it's lies or not," Damon told him, grimly. "It's going to be talked about the world over. Whoever made that tape is cashing in on the newsworthiness of the Eliminators, using their crazy crusade to ensure maximum publicity for those accusations. The tape-doctor didn't even try to make them sound convincing—he settled for crude melodrama—but that might be effective enough for his purpose, if all he wants is to kick up a scandal. Why put in those last few lines, though? Why take the trouble to include a section of tape whose sole purpose is to establish the possibility that Silas might have known his captor? What are we supposed to infer from that?"

"I don't know," Kachellek said, emphatically and defensively. "I don't understand what's happening."

"Tell me about Surinder Nahal," Damon said, abruptly.

Kachellek still hadn't recovered his usual icy calm. "What about him?" he asked, unhelpfully.

"Come on, Karol, think. Silas isn't the only one who's gone missing. Could whoever made that tape be deliberately pointing the finger of suspicion at Nahal?"

"Nahal was a bioengineer back in the old days," Kachellek said. "His field of endeavor overlapped ours to some extent. There was a little bad feeling because he thought he hadn't been given his fair share of credit for establishing the New Reproductive System, but nothing serious. I haven't heard of him in fifty years; I presume that he retired, like Silas. I can't believe that he's responsible for all this. It makes no sense. It must be someone from...."

"Someone from what?" Damon asked, sharply, when Kachellek stopped in mid-sentence again. Kachellek had no intention of finishing, though; he deliberately turned away. Whatever he knew, he wasn't going to state it aloud. Perhaps that was because he was in room whose walls might easily be host to a dozen eyes and ears, but Damon felt that it was a personal slight aimed at him, a deliberate exclusion.

"Is there any possibility," Damon said, with careful hostility, "that the viruses which caused the plague of sterility really were manufactured, by someone. Could the Crash have been deliberately caused?"

Kachellek immediately met his eye again, pugnaciously. "Of course it could," he snapped, as though it ought to have been perfectly obvious. "People don't talk about it nowadays, of course, because it's not considered a fit topic for polite conversation, but the world before the Crash was very different from the one in which you grew up. There were a lot of people prepared to say that the population explosion had to be damped down one way or another—if not by voluntary restraint, then by war, famine or plague. Primitive anti-aging technologies had already become available, and it was easy enough to see that things could get very fraught indeed when they became cheaper and more efficient. A lot of mutant viruses were arising naturally—more than ten billion people crammed into polluted supercities constitute a wonderland of opportunity for virus evolution—and a lot more were being tailored in labs for use as transgenic vectors, pest controllers, so-called beneficial fevers and so on.

"All kinds of things came out of that cauldron, as many by accident as by design. When the Crash began, speculation that it had been deliberately caused was rife; it wasn't until the Crash had almost run its course that people put that kind of talk aside to concentrate all their attention and energies on the problem of what to do about it. This is just a resurrection of ancient and tired rumors, Damon. I don't know whether to be glad or alarmed that there are so few people around who remember the last time they did the rounds. The fact remains that we didn't do it. We're not guilty of any wrongdoing."

Damon knew that Karol Kachellek had been born in 2046; he had been fifteen years younger than Conrad Helier and seven years younger than Silas Arnett. Surinder Nahal must have been much the same age—about thirty years off the current world record for longevity—but the fact that the slanders were old didn't mean that the slanderer had to be equally old. The history of the twenty-first century was all on the Web, easily accessible to anyone who cared to dig.

"Were you actually present when my father died, Karol?" Damon asked, quietly.

"Yes I was. I was by the side of his hospital bed, watching the monitors. His nanomachines were at full stretch, trying to repair the internal damage, but they just weren't up to it. He'd suffered a massive cerebral hemorrhage and there were more complications than I can count. We call ourselves emortals, but we're not really immune to disease and injury, even if we exclude the effects of extreme violence. There are dozens of potential physiological accidents with which the very best internal technology is impotent to deal even today. Kids of your generation, who feel free to take delight in savage violence because its effects are mostly reparable, are stupidly playing with fire. In essence, your father died of a massive stroke—but if the lunatic who made that tape intends to build a case on the seeming implausibility of that cause of death he's barking up the wrong tree. If Conrad had wanted to fake his death, he'd have chosen something far more spectacular."

"How did you know he was dead?" Damon asked.

"I told you," Kachellek replied, with ostentatious patience. "I was watching the monitors. I also watched the doctors trying to resuscitate him. I wasn't actually present at the *post mortem*, but I can assure you that there was no mistake."

Damon didn't press the point. If Conrad Helier had faked his death, Karol Kachellek would surely have been in on the conspiracy, and he was hardly likely to back down now. "I'm going back to L.A. as soon as I can," he said. "Maybe you ought to come with me. The people who took Silas might have designs on you too. The police can offer you far better protection in L.A. than they can in a desolate spot like this."

"I can't go anywhere," Kachellek said, stubbornly. "I've got work to do—here."

I have work to do too, Damon thought. *I know what skills it took to put that tape together, technically and in terms of its narrative implications and through Madoc I have access to some first-rate outlaw Webwalkers. I can get to the bottom of this, if I try hard enough, no matter how insistent you are on keeping me out of it.*

* * * * * * *

The last scheduled flight from Honolulu was due out at nine. Karol Kachellek—who was an important man on Molokai—had no difficulty at all in requisitioning a light aircraft and a pilot to fly it out of Kaunakakai. Damon wasn't in the least flattered that his foster-father took the trouble; he could see how keen Kachellek was to see the back of him.

Damon would rather have sat up front in the cockpit of the plane but he wasn't given the choice. He was ushered into one of the four passenger seats by the pilot, who was a stocky man with graying temples and an Australian accent. Presumably he thought the grey made him look more dignified; he looked as if he had money enough to have corrected it if he'd wanted to do so.

"We'll be up and down in no time at all," the pilot told him, before taking his own seat. "Might be a little rough in the wind, though—better keep your belt on."

Damon thought nothing of this instruction to begin with, and he was so deep in thought that ten or twelve minutes had elapsed before he finally took note of the fact that the sun—which should have been dead ahead of the plane's course—was way over to starboard.

"Hey!" he called to the pilot. "What's our course?"

The pilot made no reply. That was when Damon tested his safety-harness and found that it was locked tight. He called out again, demanding to know what was going on, but the pilot wouldn't even turn around to look at him. He realized, cursing himself for not having done so before, that he was being kidnapped.

Damon's knowledge of the local geography was vague, but he figured that if they were heading south they'd be over Lanai at much the same time that they ought to have been coming down at Honolulu. How many other islands there might be to which they might be headed he had no idea, but there were probably several, and the plane was small enough to land on any kind of strip.

"Was this Karol's idea?" he shouted to the pilot. "Or are you working for someone else entirely?"

He wasn't surprised when there was no reply. He had nothing but his powers of reason to aid him, and he didn't have enough data to work with. Furious thought merely served to multiply the questions facing him. If Karol Kachellek had instructed the pilot to kidnap him, what motive could he possibly have? Was he trying to hide Damon away, either to keep him from harm or to keep him from asking awkward questions? If the pilot was working under orders from elsewhere, who might be interested in kidnapping him? The people who already held Silas Arnett? If so, why? Did they think he had information which could be used to supplement what they had supposedly winkled out of Silas? Did they intend to force some kind of confession out of him by similar means?

It was difficult to be patient, but in the end there was no alternative. The journey wasn't significantly longer than it should have been but they overflew Lanai and headed for a much smaller island beyond it, dominated by a single volcanic peak. The plane came down just as the sun had begun to slide beyond the horizon.

When the pilot came back to release Damon from the trick harness he was carrying a gun: a wide-barreled pepper-box calculated to inflict widespread but superficial injuries. Were it to go off, Damon would lose a lot of blood very quickly, but his nanomachines would be able to seal off the wounds without mortal damage being done.

"No need to worry, Mr. Hart," the stout man said. "Nobody means to do you harm. You'll be safe here until the carnival's over."

"Safe from whom?" Damon asked. "What exactly is the carnival? Who's doing all this?"

He wasn't surprised when he received no answers to any of these questions.

The air outside the plane seemed oppressively humid. Damon allowed himself to be guided across the landing-strip to a jeep parked in the shadow of a thick clump of trees. The man waiting in the driving seat was as short as the pilot but much slimmer and—if appearances could be trusted—much older. His skin was the kind of dark coffee color which most people who lived in tropical regions preferred. He wasn't carrying a gun.

"I'm sorry about this, Mr. Hart," he said, "but we weren't sure that we could persuade you to come here of your own accord. Until we can get to the people who have Arnett, everyone connected with your family is in danger." To the pilot he said: "You can go now. Take the plane down to Hilo, just in case."

"Who are you?" Damon demanded.

"Get in, Mr. Hart," the old man said. "My name, for what it's worth, is Rajuder Singh. I knew your father and your foster-parents, long ago, but I doubt that any of them ever mentioned my name. Karol Kachellek still keeps in touch."

"Did Karol arrange this?"

"It's for your own protection. Please get in, Mr. Hart."

Damon climbed into the vehicle. The jeep glided into the trees and was soon deep in a ragged forest of thin-boled conifers. The forest was very quiet, after the fashion of artificially-regenerated forests everywhere; the trees, genetically engineered for rapid growth in the unhelpful soil, were not fitted as yet to play host to the rich fauna that ancient natural forests had once entertained. A few tiny insects splashed on the windshield of the jeep as it moved through the gathering night but there was no sound of birdsong. The road was rough and far from straight, but the driver evidently knew it well.

"Did Karol Kachellek instruct the Australian to bring me here?" Damon asked, again.

"Yes he did," Singh said, blandly. "He had to make a decision in a hurry—he didn't expect you to come to Molokai. Our people will bring the situation under control in time, but things have moved a little too fast. I'm afraid that you're in more danger than you know, Mr. Hart—I'll show you why in a few minutes' time."

"Who, exactly, are *our people?*"

Rajuder Singh smiled. "Friends and allies," he said, unhelpfully. "There aren't many of us left, nowadays—but we keep the faith."

"Conrad Helier's faith?"

"That's right, Mr. Hart. You'd be one of us yourself, I dare say, if you hadn't chosen a different path."

"Are you saying that there's some kind of conspiracy involving my foster-parents? Some kind of grand plan in which you and Karol and Eveline are involved?"

"Just a group of friends and co-workers," the other replied, lightly. "No more than that—but someone seems to be attacking us, and we have to look after our own."

"You think that Surinder Nahal is attacking you?"

"We really don't know—yet. For now, it's necessary to be careful. This is a bad time, but that's presumably why our unknown adversary chose it."

Damon remembered that Karol Kachellek had been equally insistent that this was a bad time. Why, he wondered, was the present moment any worse than any other time?

The twilight was so brief that the stars were shining long before the vehicle reached its destination, which was a sizeable bungalow set in a clearing. Damon was oddly relieved to observe that it was topped by an unusually large satellite-dish; however remote this place might be it was part of the Web; all human civilization was its neighborhood.

Rajuder Singh showed him into a spacious living-room. When Damon opened his mouth to speak he held up his hand, and swiftly crossed the room to a wall-mounted display-screen. "This is the same netboard that carried Operator 101's earlier messages," he said, while his fingers brought the screen to life.

Damon stared dumbly at the words that appeared there:

CONRAD HELIER IS NOT DEAD
CONRAD HELIER NOW USES THE NAME DAMON HART
"DAMON HART" IS NAMED AN ENEMY OF MANKIND

FIND AND DESTROY "DAMON HART"
OPERATOR 101

* * * * * * *

"It was dumped shortly before you boarded the plane at Kaunakakai," Singh told Damon, when the import of the words had had time to sink in. "Karol thought you might be inclined to argue if he showed it to you there and then. He seems to think that you always do the opposite of anything he suggests."

Damon could understand why Kachellek might have formed that impression. "It's crazy," he said, referring to the message. "It's completely crazy."

"Yes it is," said the other. "I can't understand why anyone would want to attack you in this way. Can you?"

It occurred to Damon that the people he had ordered Madoc Tamlin to investigate might resent the fact—and might possibly be scared that the buying-power of Conrad Helier's inheritance might pose a greater threat to their plan than Interpol or the friends and allies of Conrad Helier himself.

"Unfortunately," Singh observed, "such slanders can sometimes take effect before convincing rebuttals can be assembled. You see why we thought it best to remove you from harm's way. I'm sorry that you've been caught up in all this—it really has nothing to do with you."

"What has it to do with?" Damon asked, his voice taut with frustration. "What are you people up to? Why is this such a bad time for all this to blow up?"

"I can't tell you that," Singh said, with a note of apology in his voice that almost sounded sincere.

"I'll find out anyway," Damon told him—but he was wary enough not to let bravado lead him to give too much away. It might be inadvisable to boast about Madoc Tamlin's capabilities to people who might be just as reluctant to be found out as Operator 101.

The words displayed on Singh's screen suddenly disappeared, to be replaced by an urgently-flashing message, which simply said: READ NOW. The system had undoubtedly been programmed with nets set to trawl the cyberspatial sea for items of a particular kind, and one of them had just made contact. "Excuse me," Singh murmured, as he moved to claim his prize.

When Singh touched the console beneath the screen the flashing words were replaced by an image of a man sitting on a perfectly ordinary chair. Damon recognized Silas Arnett. He was not under any obvious restraint, but there was a curious expression in his eyes, and both of his hands were heavily bandaged. He began speaking in a flat monotone.

Damon knew immediately that the image and the voice were both fakes, derived from the kind of template he used routinely in his work.

"The situation was out of hand," the false Arnett said, dully. "All attempts to limit environmental spoliation by legislation had failed, and all hope that the population would stabilize or begin to decline as a result of individual choice was gone. We were still winning the battle to provide enough food for everyone, even though the distribution system left seven or eight billions lacking, but we couldn't cope with the sheer physical presence of so many people in the world. Internal technology was developing so rapidly that it was obvious to anyone with half a brain that off-the-shelf emortality was less than a lifetime away, and that it would revolutionize the economics of medicine. Wars over *lebensraum* were being fought on every continent, with all kinds of weapons, including real plagues: killing plagues.

"When Conrad first put it to us that what the world needed more desperately than anything else was a full stop to reproduction—an end to the whole question of individual choice in matters of fecundity—nobody said 'No! That's horrible!' We all said 'Yes, of course—but can it be done?' When Conrad said 'There's always a way', no one challenged him on the grounds of propriety. I couldn't see how we might go about designing a plague of sterility, because there were no appropriate models in nature—how could there be, when the logic of natural selection demands fertility and fecundity?—and I couldn't envisage a plausible physiology, let alone a plausible biochemistry, but Conrad's way of thinking was quite different from mine. Even in those days, all but a few of the genes we claimed to have 'manufactured' were actually the chance products of mutation of extant genes—we had little or no idea how to go about creating genes from scratch to have entirely novel effects—but Conrad had a weird kind of genius for that kind of thing. He knew that he could figure out a way.

"I wonder, sometimes, how many other groups must have had conversations very like ours. 'Wouldn't it be great if we could design a virus that would sterilize almost everyone on earth?' 'Yes, wouldn't it—what a shame there's no place to start.' Was there anywhere in the world in the 2070s where bioengineers gathered where such conversations didn't take place? Maybe some of the others took it further; maybe they even followed the same thread of possibility that Conrad pointed out to us. Maybe Conrad wasn't the only one who could have done it, merely the one who hit the target first. I don't know—but I do know that if you'd put that kind of loaded pistol into the hand of any bioengineer of the period the overwhelming probability is that the trigger would have been squeezed.

"We didn't discriminate: we set out to sterilize everybody. Everybody in the world. And we succeeded. That's what saved the world; if the population had continued to increase and nanotech emortality had spread like wildfire through a world which was still vomiting babies from billons of wombs, nothing could have restrained the negative Malthusian checks. Famine, war and killing plagues would have run riot. As things were, famine was held at bay, the wars cooled off and the killing plagues were countered one by one. What happened in the last three decades of the twenty-first century wasn't a tragedy at all—but the fact that it was seen as a tragedy, and a terrible threat to the future of the species, increased its beneficial effects. The Great Plague was a common enemy, and it created such a sense of common cause, focused on the development of artificial wombs and the securing of adequate supplies of sperms and ova, that for the first time in history the members of the human race were all on the same side.

"We're still living on the legacy of that break in history, in spite of attempts made by madmen like the Eliminators to set us all at one another's throats again. We're still all on the same side, all engaged in the same ongoing quest—and Conrad Helier did that. You have no conception of the debt which the world owes to that man."

"You don't regret what you did, then?" asked a whispery voice from off-stage.

"No," said Arnett's simulacrum, dispiritedly. "If you're looking for some sign of repentance, forget it. What we did was necessary, and right."

"And yet you've kept it secret all these years. When you were first accused of having done this, you denied it. When you realized that further denial was useless, you attempted to take sole responsibility—not out of pride, but out of a desire to protect your collaborators. The truth had, in the end, to be extracted from you. Why is that, if you aren't ashamed of what you did?"

"Because there are people in the world like you. Because the world is overfull of people whose moral horizons are narrow and bleak. For every man who would have understood our reasons, there would have been half a hundred who would have said 'How dare you do this to me? How dare you take away my freedom of self-determination, even for the good of the world?' Too many people would have seen sterilization as a theft, as a loss of power. Many young people, born into a world of artificial wombs, find it faintly repulsive that women ever had to give birth, but too many members of the older generations still feel that they were robbed, changed without their consent. Karol Kachellek and Eveline Hywood are still doing important work; they never wanted to be sidetracked by the kind of publicity the revelations which you've forced out of me would generate—will generate, I suppose."

"What right did you have to make decisions for all mankind?" the second synthetic voice asked, still maintaining its stage-whisper tone. "What right did you have to play God?"

"What gave us the right," Arnett's image replied, the voice as relentlessly dull as it had been throughout, "was our understanding. Conrad had the vision, and the artistry required to develop the means. The responsibility fell to him—you might as well ask what right he had to surrender it to others, given that those others were mostly ill-educated egomaniacs whose principal short-term aim was to slaughter their neighbors. Someone had to be prepared to take control, or the world was doomed. When you know that people won't accept the gift of their own salvation, you have only two choices: to force it on them, or leave them to destruction. It was better for the world to be saved—and it was better for the world to believe that it had been saved by a fortunate combination of miracles rather than by means of a conspiracy. Conrad always wanted to do what was best for the world, and keeping our actions secret was simply a continuation of that policy."

"What of the unhappiness caused by the frustration of maternal instinct?" asked the interrogative voice, in a voice devoid of any real indignation. "What of the misery generated by the brutal wrench which you administered to human nature? There are many—and not merely those who survived the Crash—who would argue that ours is now a perverted society, and that the reckless fascination with violence which is increasingly manifest in younger generations is a result of the perversion of human nature occasioned by universal sterilization."

"The empire of nature ended with the development of language," the fake Arnett replied. "Ever since then, human beings have been the product of their technology. All talk of human nature is misguided romantic claptrap. The history of human progress has been the history of our transcendence and suppression of the last vestiges of instinctive behavior. If there was any maternal instinct left in 2070, its annihilation was a good thing. To blame any present unhappiness or violence on the loss or frustration of any kind of genetic heritage is stupid and ridiculous."

There was an obvious cut at this point. The next thing Arnett's image said was: "Who told you about all this? It can't have been Karol or Eveline. Somebody must have put the pieces together—somebody with expert knowledge and a cunning turn of mind. Who?"

"That's of no importance," the other voice said. "There's only one more matter that still needs to be determined, and that's the identity which Conrad Helier adopted after faking his death. We have reason to believe that he reappeared in the world after an interval of some twenty-five years, having undergone extensive reconstructive somatic engineering. We have reason to believe that he now uses the name Damon Hart. Is that true, Dr. Arnett?"

"Yes," said a voice which sounded like Arnett's, ringing false because his head was bowed and his lips hardly moved. "The person who calls himself Damon Hart is really Conrad Helier. It's true."

* * * * * * *

Damon heard the sound of the helicopter before Silas Arnett's image faded from the screen, and immediately rounded on his companion. Singh had heard it too, and he was seized by sudden alarm. He backed away, and reached for his jacket pocket. He began to say something. The expression on his face suggested that it would be something reassuring, but Damon wasn't about to be reassured. He didn't know for sure whose side Rajuder Singh was on, but he wasn't prepared to take it for granted that it was his own.

Before Singh had any chance to say what he intended to say or to grip whatever it was he had reached for, Damon was on to him. The blow he aimed with the edge of his right hand was delivered with practiced efficiency. The old man went down with a sharp gasp of surprise.

Damon pinned him to the floor and put his own hand into the other man's pocket. He pulled out a tiny dart-gun. It was incapable of inflicting any lethal injury but it could have paralyzed him for several minutes before his internal technology mobilized itself to cancel out the effects of whatever toxin the darts bore—long enough for Singh to have made his escape from the house, if that was what he had intended to do.

Singh pried his right hand loose and tried to grab the gun, wailing: "You don't understand!"

Damon lifted the weapon out of his captive's reach but didn't hit him again; he couldn't be sure that the man bore him any ill will. "Damn right I don't," he muttered, through clenched teeth.

The noise of the helicopter was deafening now. It couldn't land because the clearing wasn't big enough but it was hovering close to the house. Damon presumed that it was unshipping men, who would burst in at any moment—but whose men would they be?

"Who are you really working for?" Damon demanded of Rajuder Singh, making his voice as harsh as he could. "Tell me now, or I'll cut you up so badly it'll take all the nanotech you've got six weeks and more to put you back together, *old man*."

Singh must have known something of Damon's past, and something of his reputation. His eyes flickered wildly from side to side, as if in search of inspiration. Damon produced the knife he always carried in his boot, for old time's sake. It had a doubly-serrated edge, designed to tear flesh in the ugliest possible way. He stroked Singh's cheek with it, and watched the blood fountain out.

"I can take your eyes out before they get here," Damon said. "And if by chance they aren't the cops, I can do a lot worse."

"It's not what you think!" the slender man gasped, seemingly desperate to spit the words out. "I really am with Karol and your father! Truly I am. If that's the enemy, you have to...."

Damon didn't find out what Singh would have wanted him to do if it had been the enemy—whoever "the enemy" might be—because the windows imploded with a deafening roar and two gas-grenades came bouncing across the carpet, pumping smoke.

"Oh shit," Damon said, lifting his arm reflexively as if to shield his nose and mouth from the fumes. He knew that it would do no good; this wasn't the first time he had been gassed by the cops. He shut his eyes tightly but he knew that it was going to sting horribly anyway, and he wasn't in the least comforted by the thought that the men outside were probably doing it in the hope of saving him from coming to any harm at the hands of his captors.

Bitter experience told him to hold off as long as possible and then to take a good deep breath, but it wasn't easy to persuade his reflexes to fall into line. He suffered several seconds of severe discomfort before he was finally able to let go and fall unconscious.

* * * * * * *

When Damon woke up he knew by the muted roar of the engines that he was aboard an airplane—not some glorified box-kite like the one the Australian had piloted, but a real intercontinental jet. He found that he was stretched out across three seats in the first-class compartment. Hiru Yamanaka was sitting on the opposite side of the aisle, watching him solicitously.

The Interpol man waited politely and patiently for Damon to gather himself together. "I'm sorry about the gas, Mr. Hart," he said, eventually. "We didn't realize that you had the situation under control, and we didn't know who was holding you. When we saw the message that was dumped immediately after your plane took off we feared the worst. Did you see it, by any chance?"

"I saw it," Damon said, sourly. "Have you interrogated Rajuder Singh yet?"

"Not yet. We confirmed his identity easily enough, but he'll be out for a long while. He seems to have had a heart attack. Perhaps you frightened him. His internal technics will pull him through, but they won't let him wake up for a couple of days. There's nothing we can do about that without imperiling his life."

Damon accepted a bottle of mineral water from Yamanaka's perennial sidekick, but waved away the offer of plastic-packaged food. He sipped slowly from the neck of the bottle, but didn't immediately raise himself to a sitting position. His head was still aching—but his mind was working well enough to alert him belatedly to the significance of what Yamanaka had said about the message advertising his peril.

"Did you say that the notice naming me was posted after the plane took off?" he said, to make sure.

"Almost immediately afterwards. That's why we feared that the two events were linked."

"Singh claimed that it was dumped beforehand—that Karol told the pilot to take me to...where was I?"

"Kahoolawe."

"Whatever—that Karol told him to take me there because he'd seen the message. If he knew about the message before it was dumped...." He stopped abruptly, wondering how best to proceed. One of the legacies of his checkered past was a deep-seated reluctance to tell the cops anything he didn't actually have to, and he still harbored the ambition of getting to the bottom of the matter before Interpol did.

"That's very interesting," the Interpol man said. "There's no evidence of Singh's involvement with the Eliminators, incidentally, or anyone else of a criminal disposition. In fact, his record is unblemished to a degree that's rather remarkable in such an old man. He's an ecological engineer, and has been for well over a century. He knew your father, although that was a long time ago."

"How did you get on to him so quickly?" Damon asked.

"We were keeping a close watch over you, Mr. Hart, even before Operator 101's third message went out. We were already tracking the plane. Did they but know it, Mr. Kachellek and Mr. Singh had no chance of stealing you away unobserved."

Did they but know it, Damon echoed, silently. *The trouble is that it's impossible to figure out how much they do know, and what their real purpose might be.* He knew that it was possible for internal technology to fake medical emergencies as well as taking action to solve them, but he didn't know whether it was unduly paranoid to suspect that Rajuder Singh had done some such thing in order to avoid—or delay—having to answer awkward questions.

"Did you catch the pilot?" Damon asked Yamanaka.

"Alas, no. The plane landed at Hilo on full automatic. He must have baled out. We're pursuing our investigations on Hawaii and Oahu, but the situation there is very confused because of another incident."

"What incident?" Damon asked, warily.

"An explosion aboard the *Kite*. Rescuers have picked up a dozen survivors so far, but there's no sign of Karol Kachellek. That's a pity—we'd hoped to ask him a few more questions about this business."

Either the unknown enemy is stepping up the violence of his campaign, Damon thought, *or....* It wasn't easy to find words to couch the alternative. "You seem to have a talent for losing geneticists," he observed, dryly. "Is Eveline still where she's supposed to be, out in L5?"

"I believe so. For the moment, I'm more concerned with the whereabouts of Silas Arnett and the identity of the persons who have been broadcasting the messages. It may not mean anything, but we've received communications from someone who claims to be the real Operator 101, disowning all the recent notices posted under that alias. It's rather difficult to check his story, of course, as he insists on maintaining his anonymity."

"I saw the tape of Silas's supposed confession," Damon said. "It wasn't him, you know—the whole thing was a fake worked up from a template. It wasn't even a particularly slick job. I could have done it better. In fact...."

He stopped, not wanting to put ideas into Yamanaka's head—but they were already there.

"In fact," the Interpol man said, smoothly, "it was a painfully obvious fake, especially at the end. Which raises interesting questions about the whole series of broadcasts. If they're supposed to look like fakes, what are we being tempted to believe, and why? Given that it's been made so very obvious that this whole case is trumped up, might we not consider it more seriously than we would had it been more expertly compiled?"

Yamanaka didn't mention the possibility—which had occurred to Damon while he watched the last tape—that the various messages had been put out from more than one source: that the third and fourth had been put out with the intention of discrediting the first and second by piling up lies and confusions. The fact that the man from Interpol didn't mention it didn't mean, however, that he wasn't aware of it; he had used the plural when talking about persons who had put out the messages.

"What did your DNA analyses tell you?" Damon asked, gruffly, as he sat up gingerly, touching his fingertips to his forehead. "You must have the results by now—and I'll bet you weren't content with superficial tissues, either. You probably drained some spinal fluid, maybe even bone marrow. What's the verdict? Am I my father, or my father's son?"

"We're completely satisfied that you're not Conrad Helier," Yamanaka told him, serenely. "If the records can be trusted, you and he have exactly the degree of genetic similarity that would be expected were you father and son. There's some uncertainty, of course, as to whether the data relating to your father's genome has been rigged to give that impression—but that kind of data is routinely filed in so many places that it would have been exceedingly difficult to alter them all. We also have a detailed record of your childhood and adolescence—it would have taken a great deal of effort to fake all that."

"It's not faked," Damon assured the policeman. "I remember it quite clearly. I didn't spring into the world full-grown. In this instance, the transparent lie really is just a transparent lie, not a cunningly-wrought truth."

"I assume, then, that we must construe it as a provocative move of some kind," Yamanaka said, evenly. "It might have been intended to startle a reaction from someone. Perhaps your kidnapping was that reaction—but it's also possible, I suppose, that it was part of the provocative move. What do you think, Mr. Hart?"

"What I think," Damon said, "is that your coming to my apartment in person was a provocative move on your part. You wanted to set me off, didn't you? I suppose you're delighted with the success of your strategy—I wasn't involved before, but I'm certainly involved now."

"You credit me with too much cleverness," Yamanaka said, with a modesty that was surely feigned. "I had no idea then whether you were involved or not, and I'm still not sure. I don't know where Karol Kachellek fits in, or Surinder Nahal—or Madoc Tamlin."

"Madoc Tamlin?" Damon echoed, trying to conceal his dismay. Clearly the Interpol man was no fool—but what was he trying to imply now?

"He's been asking a lot of questions," Yamanaka observed. "He's using your money to buy the answers. When you saw the second message you called him and you extended his authorization. You presumably believe that you're pulling his strings—but I have to consider other possibilities too. I have to consider the possibility that your strings are being pulled. This is a very convoluted puzzle, Mr. Hart, and Madoc Tamlin has some very convoluted friends."

"So have you, Mr. Yamanaka," Damon countered.

The man from Interpol didn't deny it. Instead, he said: "Dr. Arnett's supposed confession was an interesting statement, wasn't it? Food for thought for everyone—and food which will be all the more eagerly swallowed for being dressed up that way. I dare say that he was right about the effect the Crash had, of bringing people together so that for the first and only time in human history they were all on the same side. The world isn't like that any more, is it? In a way, that's rather a pity, don't you think?"

"Not really," Damon replied. "A world devoid of conflicts would be a very tedious place to live. It's good to know that we might live for a very long time—but it's also good to know that we might not. Without an element of danger, life might easily become insipid." He felt a lot better now, and he was able to sit up.

"I take your point," Yamanaka conceded, graciously, "but you must remember that you and I are young men, who can barely imagine what the world was like before and during the Crash. I wonder, sometimes, how different things might seem to the very old—to men like Rajuder Singh, Surinder Nahal and Karol Kachellek, and women like Eveline Hywood. They might be rather disappointed in the world they made, and the children they produced from their artificial wombs, don't you think? They were hoping to produce a Utopia, but...well, no one could convincingly argue that the meek have inherited the world—at least, not yet."

Damon didn't know what Yamanaka might read into any answer he gave, so he prudently gave none at all. *He just wants to use me as a pawn*, he thought. *Maybe I could have stayed out of it, if I hadn't gone to Madoc the minute he alerted me to what as going on, but I'm in now for better or for worse, and I have to play it through—not for Interpol, and not for my father's true blue friends, but for myself.*

"Sometimes," Yamanaka added, in the same off-handedly philosophical tone, "I wonder whether anyone *can* inherit the world, now that people who owned it all in the days before the Crash believe that they can live forever. I'm not sure that they'll ever let go of it deliberately...and such fighting as they have to do to keep it is mostly amongst themselves."

"My father never owned more than the tiniest slice of the world," he said, awkwardly conscious of the fact that he had said *my father* instead of *Conrad Helier*. "He was never a corporation man."

"Your father remade and reshaped the world by designing the New Reproductive System," Yamanaka replied, softly. "The corporation men who owned it might well have hated him for that, even though he never actually succeeded in toppling their commercial empire. Men of business always fear and despise Utopians. They probably hate him still, almost as much as the Eliminator diehards hate them."

"But he's been dead for fifty years," Damon pointed out. "Corporation men wouldn't waste time demonizing the dead."

"His collaborators are still alive," Yamanaka countered. "Or were, until this plague of evil circumstance began."

* * * * * * *

When the plane landed in Los Angeles Damon was invited to accompany Hiru Yamanaka and his associate to the local Interpol headquarters, but he declined. Despite stern warnings regarding Interpol's inability to guarantee his safety, he insisted on going back to his apartment—and in the end, Yamanaka agreed to take him there.

"The claims made by the so-called real Operator 101 are, of course, receiving a full measure of publicity," the policeman told him. "They have not gone uncontradicted, but would-be assassins might be not inclined to believe the contradictions. You really would be safer in another location."

"You can't take me into custody," Damon said, obstinately. "I haven't done anything wrong. If I thought I needed bodyguards, I could hire some very experienced street fighters."

"That would be unwise," Yamanaka said, blandly. "My advice is to leave Madoc Tamlin and your former friends out of this. They're essentially unreliable."

Damon had his own views on that particular matter, but he didn't try to recruit any armed guards of his own and he didn't object when Yamanaka's taciturn companion didn't get back into the car when they dropped him outside the capstack.

"Just a precaution," the policeman said, as they rode the elevator to the thirteenth floor. "I won't camp outside your door, but I'll be around."

Damon knew how easy it was to mount eyes and ears in the walls of the corridor, and he didn't doubt that anyone approaching his apartment would be under constant surveillance. Yamanaka hadn't made any false promises about respecting his privacy.

When he'd taken time out to visit the bathroom and order some food from the kitchen dispenser Damon stationed himself before the windowscreen. He wasn't unduly surprised or alarmed when Madoc Tamlin's phone insisted that he was unavailable. He half-expected to get the same response from Eveline Hywood, but in fact she answered immediately. She even came on camera, so that the time delay occasioned by the fact that their words and gestures had to traverse a quarter of a million miles wouldn't be quite as disconcerting.

"Damon," she said, pleasantly. "It's good to see you. I've been worried about you. Is there any news of Karol or Silas?" She was obviously well-informed about what had been going on.

"They haven't been found yet—dead or alive," he told her. "Interpol's man insists that it's only a matter of time. Do you have any idea what's going on, Evelyn?"

"Someone is evidently intent on blackening your father's name. I can't imagine why. These self-appointed Eliminators seem to be getting completely out of hand. There are none up here, mercifully; L-5 isn't perfect, but it's a haven of perfect sanity compared to Earth."

Damon didn't bother to question her certainty as to whether L-5 was really Eliminator-free. For the moment, he was inclined to the opinion that the aggrieved Operator 101 really was the victim of a pseudonym-hijack and that this whole affair was a struggle between two very different groups.

"Why now, Eveline?" he asked, softly. "What brought your adversaries crawling out of the woodwork now?"

"I have no idea," she said. He couldn't tell whether she was lying. "You might be better able to guess than I am. After all, this whole affair is really an attack on you, isn't it?"

It is now, he thought. *But it didn't start that way. That's a deflection, a diversionary tactic, for which my father's so-called friends are at least partly and perhaps wholly responsible.*

"Could it have something to do with this stuff that you and Karol are investigating—these para-DNA life-forms?" he asked, abruptly. That was the only thing that was happening now, so far as he could judge—the only thing that made it a "bad time".

"How could it?" she asked, frowning. Was she puzzled, he wondered, or annoyed by the accuracy of his guess?

"Karol said there were two possibilities regarding its origins: up and down. He was looking at the bottom of the sea while you're looking for evidence of its arrival from elsewhere in the solar system. But he seemed to have a third alternative in mind when he said it—and there is a third alternative, isn't there?" He knew that he didn't have to spell it out that the third alternative was sideways; Eveline understood well enough what he meant.

"I'm still very worried about you, Damon," Eveline said, scrupulously ignoring his question. "I wish you were safe. I'm sure it will all work out, though, if you only give it time. When they find Silas, he'll put the record straight."

"What about Surinder Nahal?" he asked. "Could he really be the one behind this stupid carnival? Does he really think my father orchestrated the Crash as well as the recovery? Why hasn't he said so before?"

"I don't know, Damon," she said, with exaggerated patience. "It's all lies. You know that."

"Is there going to be a new plague?" he asked, abruptly switching tack again. "Is para-DNA going to throw up something just as nasty as the old meiotic disrupters and chiasmalytics?"

"That's ludicrously melodramatic," she answered, calmly. "So far as we can tell, para-DNA is quite harmless. Organisms of this kind compete for resources with life as we know it, but there's no evidence of any other kind of interaction and it would be surprising if there were. Para-DNA is just something which happened to drift into the biosphere from elsewhere—probably from the outer solar system. It's fascinating, but it's not dangerous."

"Are you sure it came from the farther reaches of the solar system?" Damon asked, determined not to let the matter lie.

"Not yet," she answered, equably. "Investigations of this kind take time, and we have to be very careful to have all the data in place before we draw conclusions."

"Yes," Damon said, in a neutral tone which was meant to imply far more than the words could say. "I understand that." He really thought he did. That part, at least, he thought he had figured out.

"Please be careful, Damon," Eveline said. "In spite of our past disagreements, we all love you. We'd really like to have you back one day, when you've got all the nonsense out of your system."

I believe you would, he thought. *In fact, I believe you think you will*. All he said out loud was: "I'll be careful. Don't worry about me. You've got better things to do."

He tried again to reach Madoc Tamlin, but he failed. He still didn't read anything sinister into that, until he received an incoming call from Hiru Yamanaka informing him that Diana Caisson had been arrested in Oakland, and that she was being sent back to Los Angeles for questioning about her possible implication in the suspicious death of Surinder Nahal.

* * * * * * *

At first, Yamanaka refused to let Damon talk to Diana, although he admitted that she had asked repeatedly to see him. The Interpol man also refused to discuss the details of the case which he was supposedly building against her, although he confirmed that she had been captured while fleeing with a companion from a house where Nahal's body had been found, and that the police were still searching for Madoc Tamlin, who had been conclusively identified as the companion in question. It wasn't until Interpol received confirmation from the Oakland police that the body discovered in the house had been dead for some considerable time before Diana and Madoc Tamlin had arrived there that he relented.

"Will you let her go now?" Damon asked, as he was taken down to the holding cells beneath Interpol's L.A. headquarters.

"I can hold her for a while longer," Yamanaka told him. "I'll charge her with illegal entry if I have to. I'd like to talk to your friend Tamlin before I let her go, if only to cross-check her claim that she doesn't know why they went there. I do wish you hadn't involved Tamlin in this business; it's an unhelpful complication. When you offer money for information you attract all manner of spiders—not just the clever crackers who spend all their real-time poking around the Web but the poisonous ones who prey on anyone and everyone."

"He seems to have located Surinder Nahal before you did," Damon pointed out. "Did Madoc lead you to him, by any chance?"

"No, he didn't," Yamanaka replied, in a faintly offended tone.

"You were acting on information received, weren't you?" Damon guessed. "The people who tipped you off were the ones who got to Nahal before Madoc did. They're one step ahead of all of us, aren't they? Do you have any real idea yet who they are?"

"Some of them are one-time friends of your father's, Mr. Hart," Yamanaka said, perhaps feeling the need to demonstrate once again that he wasn't a fool. "I think they knew that Silas Arnett had been kidnapped before we began looking for him, but that they preferred to try to handle things on their own—just as you did. Independence of thought and action seems to run in the family—and the Eliminators are far from being the only secret society supported and sustained by the Web."

There was no time for further talk; Yamanaka let him into the cell where Interpol's agents were keeping Diana. She seemed to be glad to see him, even though she hadn't forgiven him anything. There was a noticeable tension in their embrace.

"This is crazy," she said. "They must know we didn't kill the guy. We didn't even know the body was there."

"They know you didn't kill him," Damon reassured her. "What on earth possessed you to go there? Why was Madoc fool enough to let you?"

"He asked me to help him," Diana said defensively. "He didn't tell me what he was doing. He just wanted me to talk my way into the place—spin a line to persuade the guy to open his door. It wasn't necessary; the door was open when we got there. We didn't even have time to start stripping data from the guy's systems. The cops must have done that after they picked us up; whatever there was, they've got. Can you get me out, Damon? You owe me that much."

While she talked, Diana moved her hands nervously back and forth. Damon didn't doubt that the walls had eyes which could see every last gesture, but he was fairly sure that the patient watchers wouldn't be able to decode the signals she was sending. In a world where any environment might be bugged, people like Madoc Tamlin were careful to develop private codes of communication, known only to their closest friends. Diana clearly wasn't an expert in the use of this one, but she knew enough to spell out a name, repeating it to make sure he got it.

The name she was signing was "Lenny". There was only one Lenny she could mean.

"I'll get you a lawyer," Damon promised. "I don't think they're really interested in pressing charges—they just need an excuse to talk to Madoc for a while. They want to know what he found out, just for curiosity's sake. It'll be okay. You'll be out in no time."

As expected, Yamanaka was waiting for him outside.

"What did you find out?" the policeman inquired, politely.

"Nothing you didn't overhear," Damon assured him, not expecting to be believed. "How are things going at your end?"

"As your girl-friend said, we stripped the data from the systems in the house in Oakland and we're going through it with a fine-toothed comb. It's possible that Arnett was held there, but he's certainly not there now and we can't be sure. The fact that Nahal died in problematic circumstances gives us carte blanche to root through everything he left behind—if he is behind this puppet-show, we'll uncover every last detail of it eventually. It's just a matter of time."

Damon was alert enough to note the peculiar circumlocution. "What do you mean, *problematic circumstances?*" he asked. "I thought he'd been murdered."

"According to the medical examiner," Yamanaka said, "he wasn't—not directly, at any rate. He seems to have died of natural causes. The only mystery is why his internal technology didn't prevent it, but it's probable that he was simply too old. We're so used to nanotech magic that we've come to expect miracles which even the cleverest machinery can't deliver."

My father died of natural causes too, Damon thought, *and I dare say that poor Karol will turn up drowned. Rajuder Singh hasn't recovered consciousness yet, and perhaps he never will. If Silas is dead, too, that only leaves Eveline unaccounted for. Except, of course, for me.*

"We live in a very complicated world," Damon said, matching the Interpol man's oddly irritating philosophical manner. "We're so good at creating virtual realities that we've almost lost the trick of distinguishing appearance from reality. Maybe we expect more of social machinery like Interpol than Interpol can possibly be expected to deliver."

"Are you talking about truth, or justice?" Yamanaka countered.

"Both," said Damon, dryly.

* * * * * * *

When he left Interpol headquarters Damon immediately headed for the most crowded streets in the city. He was reasonably sure that Yamanaka's taciturn companion was still looking out for him, and that he wouldn't be easy to shake off. He bought a new suit of clothes and left the old one behind, just in case Yamanaka had planted any discreet bugs on his person, and he stopped off at a public gym for a shower, just in case there was anything in his hair that shouldn't have been there. His internal technology was good enough to take care of anything that had got under his skin. He looked up Lenny Garon's address on the gym's directory-terminal.

When he left the gym he struck lucky. A software glitch put half the local traffic signals out of action for a full five minutes—time enough to snarl up twenty thousand vehicles and create a jam which would require at least an hour to clear. The pavements jammed up almost as badly as the gridlocked vehicles, and tempers soared all along the line. Damon kept on ducking and dodging until he was certain that he was free and clear of all humanly possible pursuit, and then began the painstaking business of making his way across town without leaving a Webtrack.

"Is it about my tape?" Lenny Garon said, anxiously blinking his one good eye, as he let Damon into his squalid capsule. "Did something go wrong with the mesh?" Tamlin plainly hadn't let him in on any secrets.

"Your tape's fine," Tamlin reassured him, once Damon was safely inside. "Just take a walk, will you. The two of us have something private to discuss. I'll pay you a couple of hundred in rent, but you have to forget you ever saw us, okay?"

The kid was appropriately impressed. "Be my guest," he said. As he disappeared into the corridor, he called back: "I hear you're an enemy of mankind now. Good going."

As the door slid shut behind the boy Damon looked around the room, wondering that people still had to live like this in a world whose population explosion had fizzled out long ago. While the greater part of L.A. slowly rotted down into dust its poorer people still huddled together in neighborhoods full of high-rise blocks full of narrow rooms with fold-down beds, kitchens the size of cupboards and even smaller bathrooms. Perhaps people had grown over-accustomed to crowding during the years before the Crash, and now couldn't live without it; that made more sense, in a way, than conventional explanations about buildings needing services and the proximity principles of supply and transport.

"What the hell is happening, Damon?" Tamlin asked, when the door was firmly closed and Lenny Garon's footfalls had died into silence.

"You tell me."

Tamlin shook his head again. "Damned if I know. How bad is the trouble I'm in?"

"Nothing much. They know Nahal was dead before you got there. All you did was find a body. They'd be grateful if it weren't for the fact that somebody else had already tipped them off about it. They still want to talk to you, but they'll be polite."

Tamlin's relief on hearing this news was very evident. "Who killed him?" he asked.

"Nobody. At the very worst, somebody flushed out his internal technology, the way Silas Arnett's kidnappers did. Perhaps it's supposed to look like justice. What did you find out, Madoc?"

"Not much," Tamlin admitted. "The way the spiders are spinning, it looks as if this guy Nahal had some kind of grudge against your father and his cronies. Maybe he'd been nursing it for a hundred years, or maybe it's just something that came back to haunt him in his old age. It looks as if Nahal had Arnett snatched, and that he put out the Operator 101 stuff himself—although the word is out that he isn't the same guy who built up the Operator 101 name and reputation. He was difficult to trace, but not too difficult. I'm sorry about Diana—she wouldn't stay home. You know how she is. I didn't tell her anything."

Damon had taken note of the emphases. "You said *looks*?" he queried.

"That's right," Tamlin confirmed. "I've no proof, but I have this itchy feeling that looks are all there is. Even before I found the door open and the guy lying dead...I think it was left all neat and tidy for someone to find—the cops, I guess. I get the impression that somebody's busy clearing up their case for them. Whoever it is, they're cleverer than any people I know, and they have more money. That's what I think, anyhow."

"It's what I think too," Damon said. "You warned me when I asked you to help that all the best outlaw Webwalkers were in the pockets of the corps. I didn't think it was relevant, but it is. All we could ever buy was a slice of the same pie they were feeding the authorities."

Tamlin shook his head, wonderingly. "What's it got to do with the corps?" he asked.

"Nothing, except insofar as the real owners and movers of the corps are hidden. The trouble with a world in which it's difficult to keep secrets is that everyone tries so much harder. The Web is an open book to those who know how to turn its keys, and nanomachinery makes all kinds of unobtrusive eavesdropping as simple as falling off a log—as well as very, very cheap. There's only one option for anyone who wants to move behind the scenes, and that's to throw up multiple smokescreens. The only way to hide the truth is to dissolve it in an ocean of bluff and double-bluff. You and I wouldn't even be clever enough to figure out who the players in this game might be, if it weren't for the accident of biology that connected me to one of them—and even that would be irrelevant if it hadn't been for the vanity which made my father instruct his trusty hirelings to do everything possible to turn me into a replica modeled on the same template."

"Why did they say you were Conrad Helier? They couldn't possibly expect people to believe that."

"They didn't. The third and fourth messages were calculatedly transparent lies, intended to discredit any truth that might have been lurking in the first two. My father's team also wanted to give themselves an excuse for taking me out of the game—or dragging me further into it, on their side rather than the opposition's. They didn't approve of my sending you out digging for information—but they cocked up the abduction because they were in too much of a hurry."

"Are you saying that your father really is alive, and that he really did cause the Crash?"

For a moment, Damon was tempted to tell his old friend exactly what he did believe—but only for a moment. When the moment was past he knew that he'd made a crucial decision.

"No," he said, after only the slightest pause. "He's dead all right, and if the viruses that caused the Crash were manufactured, he didn't do it. We just got caught in the crossfire of an old war, which should have been laid to rest fifty years ago. It really doesn't matter a damn who wins and who loses. Anyhow, Surinder Nahal has been cast in the role of guilty party, and now he's dead the case will be closed."

"You don't think Nahal did it, do you?" Tamlin said, just to be sure.

"No," Damon said, "I don't think he did. I think my father's friends want to keep the real guilty party out of sight, so that the only files that will be examined and picked apart will be Nahal's—and those will be very carefully doctored. That way, they can shape the disclosures to suit themselves." That much, he figured, was almost certainly true—but it was only a part of the truth. What the hidden movers wanted to keep hidden was something more significant than the possibility that Conrad Helier had designed and released the viruses which caused the Crash.

"Things could be worse," Tamlin opined—still, apparently, buoyed up by the relief of knowing that Interpol had no particular grudge against him. "At least the bastards are still dying, one by one. Imagine how much worse it'd be if they really could live forever. No matter how far or how fast our generation went, we'd always be one step behind them. It'll all be ours one day, though—we just have to be patient."

Tamlin had always had an uncanny knack of putting his finger on the heart of a problem, although he was perversely prone to misinterpret the significance of what he touched. Damon was about to congratulate him on his cock-eyed perspicacity, with all due irony, but as he opened his mouth he saw Tamlin's expression change to one of horror and alarm. He turned abruptly to see the apartment door sliding noiselessly into its bed.

A long arm, which was certainly not Lenny Garon's, reached around the jamb and lobbed something into the tiny room. It fell at Damon's feet. He had only seen such objects in broadcast VR dramas, but he recognized it immediately as an explosive grenade.

There was nowhere in the room to hide and the door was already sliding shut again. Had Madoc Tamlin been a hero of the self-sacrificing kind he might have tried to bundle Damon along in front of him as he headed for the exit, keeping his own body between his friend and the threat of mortal injury, but he wasn't that kind of hero—and Damon couldn't blame him for it.

What Tamlin actually did was to dive past Damon and hurl himself at the slowly-closing slit that led to relative safety. He didn't make it. He couldn't even get his fingers into the crack before the escape-route was sealed off. The doorway was just as narrow as everything else in the apartment, and the door was very efficient.

Damon watched, dumbly, as Tamlin put his arms over his head and huddled up into a quasi-fetal position, evidently hoping that his internal technology might be good enough to pull him through the effects of the explosion, if only he could make himself a small enough target.

Damon was startled by his own composure as he bent down and picked up the grenade.

"It's not real," he said, after holding it for ten or twelve seconds. "It's a fake."

Tamlin wasn't immediately convinced, but it didn't take him long to realize that he must look incredibly foolish. He slowly unwound, and looked Damon in the face. There was no relief in his expression now, although it would not have been inappropriate.

"What kind of crazy man would throw a fake grenade into a place like this?" he asked, harshly.

"The kind who wanted to deliver a message," Damon replied, dully.

Tamlin looked at the grenade, as if he expected Damon to unscrew the cap and produce a piece of paper from its hollow interior. "What message?" he asked, as he came slowly to his feet.

"It says: *If we really wanted you dead, you'd be dead*," he explained, softly. "You don't have to worry about it, Madoc—it's for me."

* * * * * * *

When Damon got back to his apartment Hiru Yamanaka was waiting for him. The Interpol agent was alone. He was sitting in a chair, looking comfortable, but Damon knew well enough that he must have spent the bulk of the time he'd been there prying, with a more ruthless efficiency than Diana had ever contrived. He'd probably found evidence of half a hundred minor misdemeanors, but he would doubtless file them away, at least for the time being.

"That wasn't very wise, Mr. Hart," Yamanaka said. "You could have put yourself in danger."

"If anyone really wanted me dead, I'd be dead," Damon told him.

"Really dead, or only apparently dead?" Yamanaka asked, innocently.

"How's the investigation?" Damon countered.

"All wrapped up," the man from Interpol said. "The evidence was delivered to us on a plate when we stripped the systems in that house up in Oakland. It seems that Surinder Nahal had suffered the fate which so many of our old men fear. His internal technology hadn't been able to maintain his brain to the same standard as his body, and he'd fallen prey to mental illness. Having belatedly conceived a paranoid hatred for his more famous contemporary Conrad Helier, he kidnapped Silas Arnett and employed a mixture of straightforward torture and deceptive videosynthesis to build an entirely false case against the man of whom he was so envious. All the evidence to prove it is now in place. We found Arnett's body, by the way—and what was left of Karol Kachellek's. Arnett seems to have died of injuries inflicted after his internal technology had been flushed out; Kachellek might or might not have drowned before the sharks tore him apart. My investigation wasn't a complete failure, however—we were able to track down the original Operator 101 when she started complaining about the usurpation of her pseudonym. I gather that she's rather looking forward to her day in court, in anticipation of being able to plead the Eliminator cause with all due eloquence before a large video-audience. I do hope the newstapes won't make a martyr of her."

"It's a perverse world we're living in, Mr. Yamanaka," Damon said. "Appearances matter far more than reality. In fact, we've sophisticated our virtual realities to the point where the distinction between appearance and reality has broken down. Each layer of illusion that we penetrate merely reveals another layer of illusion."

"You know that's not true, Mr. Hart," the policeman said. "No matter what practical difficulties people like you and I might encounter in getting to the bottom of things, there really is a bottom. The truth is there, no matter how well-camouflaged it might be. If I didn't believe that, I'd be no use at all as an investigator."

"Are you satisfied that you've reached the bottom of this business?" Damon asked him, knowing the answer already. "Have you sorted out the truth from the morass of disinformation in which it's submerged?"

"Not yet," Yamanaka replied. "But I won't stop trying. Will you?"

"Did Rajuder Singh ever wake up?" Damon asked.

Yamanaka nodded. "He told us that Karol Kachellek asked him to take care of you because he thought you were in danger, and that he knows no more than that. He says he didn't know that the Operator 101 message naming you didn't go out until after Kachellek had called him. It's probably true, and if it's false we'll never be able to prove it. Do you want him charged with false imprisonment?"

"No. Let the matter rest."

"Is that what you intend to do?" Yamanaka asked, again.

"I should never have hared off on a wild goose chase in the first place," Damon told him. "I guess I must still be unduly sensitive about the matter of my supposed birthright, and I was all strung out because of the business with Diana—but it's all over now. My father's dead, and so are most of my foster-parents. There's only Eveline left, and she's a quarter of a million miles away. Diana won't be coming back. I can get on with my own life now, and that's what I intend to do."

"You're a liar, Mr. Hart," Yamanaka said, with an off-hand calmness which didn't take the sting out of the words. "You know as well as I do that the messages I first brought to your attention weren't just raking over burnt-out embers. They were building up to some other revelation, but then your father's friends stepped in and took over the script, putting their own scapegoat—with a full complement of supplementary evidence—in place of the one the opposition intended to use ."

"I suppose they killed him just to make it look good," Damon said, to test the water.

"You know how easy it is to synthesize appearances," the policeman said. "That doesn't just apply to images transmitted over the Web and VR tapes. A first-rate biological engineer can probably fake genetic appearances as easily as a good tape-doctor can fake visual ones. There are far too many bodies in this affair, Mr. Hart—too many dead men whose internal technology ought to have kept them alive but somehow didn't. A *corpus delicti* isn't sufficient evidence of death in a world like ours—a world, where flesh can be manufactured and shaped, where the physical appearances of the living can be modified by somatic engineering and where new identities can be so easily created by throwing bits of data into the chaotic flow of the Webstream."

Damon recalled what Madoc Tamlin had said about things being better than they might be. While the old continued to die, the young still had a chance to inherit the empires of the world—so it was in the interests of the old to maintain that image of the world. It was in the interests of the people with the very best internal technology to play down its power—to maintain the idea that what people called immortality wasn't really immortality at all, or even emortality. It was in the interests of the people who owned the corps that owned the world to persuade their would-be heirs that patience was still the cardinal virtue, that their elders were liable to lose their memories and their minds and were still certain to die, in the end. But if all of that were mere appearance and mere illusion, what hope would there be for the impoverished young ambitious to claim a generous slice of the big cake?

The Eliminators offered hope of a nasty kind, but Damon knew only too well that it was a false hope, a mere colorful folly. Damon knew—and he was sure he knew it because he was his own man, and not because he was his father's son—that the one and only real hope in a world like that was of a very different kind.

"With your help," Yamanaka said, "I might be able to dig a little deeper. You're in a better position to put pressure on Eveline Hywood, and pry into her affairs, than I could ever be. Together, we might be able to penetrate this sham and figure out what the real dispute is all about. My guess is that it's about para-DNA. The people who launched this attack on your father's inner circle intended to expose the fact that it's a fake—something cooked up in a lab, just like the viruses that caused the Crash."

"You're dreaming, Mr. Yamanaka," Damon told him, equably. "You're living in a virtual reality of your own design. Para-DNA is a product of nature; my guess is that it drifted in from outer space. Anyone who said otherwise would be a liar—an obvious liar."

Yamanaka didn't scowl or shrug his shoulders; he just got up from the chair and quietly adjusted his clothing before heading for the door. Damon was expecting a Parthian shot, though, and he wasn't disappointed.

"I'm disappointed in you, Mr. Hart," Yamanaka said, as he let himself out. "I thought that you really were determined to be your own man, and to escape the tentacles of your father's schemes and ideals. I thought that you might at least have a healthy resentment of that trick they pulled in naming you. No matter how many denials are broadcast, you'll never be entirely safe from the Eliminators. Don't you resent the fact that they're still manipulating you, even after all these years."

"You know what people say," Damon countered, unwilling to let the policeman have the last word. He hesitated just long enough to conjure up a quantum of suspense and dramatic tension before adding: "If you can't beat them, join them."

* * * * * * *

The contact was so long in coming that Damon had almost stopped expecting it. Months passed: months which he spent in splendid isolation, creating imaginary worlds for anyone and everyone who would pay him to do it, making no discrimination between Madoc Tamlin's black market tapes and those commissioned by legitimate corporations. He didn't bother to keep a close watch on the news; by the time he heard that Eveline Hywood had confirmed that para-DNA was an alien invader carried into the inner solar system by comets from the Oort Cloud, such excitement as the announcement had generated was already dying down. Nobody tried to call her a liar—no one, at any rate, who could get a hearing from the people who put the newstapes together.

Appropriately enough, the contact, when it finally came, was made in the one place where Damon thought his privacy really was guaranteed: in an imaginary world that he was in the process of building. He was designing a sharespace for use in an adventure game, but it wasn't nearly ready to be opened up for sharing. It was a big commission, requiring him to design the natural phenomena, flora and fauna of a hypothetical alien world orbiting a distant sun, whose visitors might undergo all kinds of vivid adventures, individually or in groups. He had his VR-apparatus hooked up to the Web so that he could decant commercially-available templates for adaptation and integration, but nothing was supposed to be able to come down the cable unless he summoned it. When he realized that there was someone else wandering around in his creation he felt a strange sense of violation that was even more shocking, in its way, than the appearance of the stranger. The appearance was, after all, a mere fiction; although it looked exactly like the pictures he had seen of his late father, it could have been anyone at all.

"Hello, son," the image of Conrad Helier said, softly. "We meet at last." As he spoke he looked around at the multicolored alien jungle and all the vivid insects with which Damon was busy populating it; his gaze seemed slightly disapproving, as if he found it all rather tawdry and inartistic.

Damon couldn't entirely disagree with such a judgment, if it were indeed implied; he had been instructed—in so many words—to paint in gaudy and lushly unnatural colors, to think Douanier Rousseau rather than Corot or Constable.

"You people really are full of surprises," Damon said, determined to hold his own in what was bound to be a problematic discussion. "I suppose sophisticated biotechnics and clever nanomachinery are so similar to magic that you've all started behaving like the magicians of legend: jealous, secretive, loving deceit for its own sake."

"Not for its own sake," the other said, with a sorrowful shake of the head. "The opposition is secretive, and fiercely jealous of its secrets, because its power is based in products and profits, patents and petty monopolies. To people of their sort, knowledge is capital to be hoarded and guarded, invested at the highest available rates of interest."

"And you're different, I suppose?"

"Yes we are," Conrad Helier retorted, firmly. "Their end is our means. They don't have any long-term objectives except for preserving their advantages and maintaining their comforts. They only want to control things because they couldn't bear to be controlled. Even though they're effectively immortal, they're still thinking in terms of today and tomorrow. They'll grow out of it, in time—but until they do, they're a heavy anchor holding progress back. We're planners and builders. We think in terms of centuries and millennia. We're practicing to be masters of evolution, but in the meantime we're trying to be midwives of history. We're not interested in money for its own sake, or power for its own sake. We're interested in what money and power will enable us to do."

"And pretending to die was just a career move, I suppose?"

"If you want to put it that way. It wasn't very difficult. I simply manufactured a second body—a clone, if you will—complete with all faults, for the benefit of the autopsy. You'd be surprised how easy it is to contrive a simple gypsy switch, even in a hospital, proved that no one's expecting it."

"And you did the same for Karol Kachellek and Surinder Nahal. What about Silas? Did you manage to get him back in time, or did the opposition really kill him?"

"We got him back. They released him once their pitch had been ruined. That's one good thing about the way the game is being played—nobody fights to the death any more. Emortals tend to be scrupulously careful about that sort of thing."

Damon remembered the fake grenade, and he nodded. "I guess that's progress," he admitted. "It might introduce an element of farce, but it's better to play war-games than fight authentically bloody wars."

"It was no part of our plan to involve you," Conrad Helier's simulacrum said, "but once you'd involved yourself we had to treat you as a player. You do understand that, don't you?"

"I understand," Damon said. "I caught on pretty quickly, didn't I?"

"Yes, you did—quickly enough to make a father proud. Not that you could avoid your destiny forever, of course. No one can. You ought to be able to adapt more readily than most. You are my son, after all."

Damon didn't entirely like the tone of these remarks. Even now, he didn't want to be taken for granted. "Was Silas's supposedly fake confession true?" he asked, abruptly. "Did you design the viruses which caused the Crash?"

"I designed one of them. To this day, I don't know who designed them all, and it's certainly possible that some of them really did arise naturally. We didn't kill anybody, Damon—we just took away the supposed right which people claimed to multiply themselves to the point at which the side-effects of their living destroyed the ecosphere. That statement Silas quoted is perfectly accurate: we had to play God, because the position was vacant."

"And you're still doing it—but you have to move in mysterious ways, because you're not unopposed."

"Again, if you want to put it that way—yes. Somebody has to make plans, Damon. Somebody has to ask the big questions. If everybody were prepared to join in, we'd be only too happy to let them—but there are too many people in the world who only value the moment and aren't prepared to think about the more distant consequences of their actions. You understand that very well, I think."

Damon felt a surge of resentment, but he didn't try to contradict the phantom. "You designed para-DNA too, didn't you?" he asked, instead.

"I didn't do it on my own—but it is a laboratory product. We think that the Earth needs an alien invader, Damon: an all-purpose alien invader, which can turn its hand to all kinds of projects."

"Why? What's the point?"

"Because we can't afford to export our spirit of adventure to virtual reality, and there's very real a danger that we might do just that—as evidenced by the fact that you and I have to meet like this in order to avoid the millions of microscopic eyes with which the real world is dusted. People shouldn't be living in the ruins of the old world, contentedly huddling together in the better parts of the old cities and binding themselves ever more tightly to their particular stations in the Web, like flies mummified in spidersilk. Nor is it rebellion enough against that kind of a world for the disaffected young to use derelict neighborhoods as adventure playgrounds where they can carve one another up in meaningless ritual duels. We have to maintain some kind of movement, because without movement there'll be no momentum. People have to build and keep on building, to grow and keep on growing. We have to make progress, Damon—and if people need a spur to urge them on, I'm more than willing to provide it."

"And para-DNA is your spur. Another plague, like the ones the God of the Old Testament rained down on stubborn Egypt in order to secure the release of His people from captivity." Damon's voice dripped sarcasm.

"It's not nearly as devastating as the slaughter of the first-born," his father's simulacrum pointed out. "Like the viruses that caused the plague, para-DNA is no killer—but it could be a terrible nuisance. It might attack the structure of the cities and the structure of the Web; it might make it impossible for the human race to dig itself a hole and live in manufactured dreams. It won't attack people, and it certainly won't murder people wholesale, but it'll always be there: a sinister, creeping presence that will keep on cropping up where it's least expected and where it's least welcome, to remind people that there's nothing—nothing, Damon—that can be taken for granted.

"Long life, the New Reproductive system, the earth, the solar system...all these things have to be managed, guarded and guided. We ought to be looking towards the real alien worlds instead of—or at least as well as—synthesizing comfortable simulacra; my people are just trying to make sure that happens. As I said, it's a long-term plan; nothing melodramatic will happen for a few years, but we couldn't afford to have the plan aborted before we even got it off the ground. We have to keep up appearances for a long time yet. I'm sure you understand that. Why else would you turn down Mr. Yamanaka's tempting offer to turn traitor?"

"Why did the corps try to sabotage the plan?" Damon wanted to know. "It sounds to me as if you'll be stimulating a lot of economic activity"

"The corporations don't like us. They really would prefer it if the meek inherited the earth. The corporations are only interested in what people want, and the more stable and predictable those wants are, the better the corporation men like it. We're interested in what people need, and that makes us difficult to figure. They don't think of us as competition, because we're not aiming for profit and our risk calculations aren't made by accountants, but we're an irritating thorn in their side nevertheless. They're not about to launch an all-out crusade against us, because it wouldn't make economic sense, but they're always prepared to make a small investment in a good spoiling tactic. They probably figure that they've won a tiny victory by forcing a few more of our personnel into effective invisibility, but that just goes to show what small-minded cowards they are. In the end, they can't win—because they're not really playing to win; they're only playing for money."

"Playing God must be addictive," Damon observed, neutrally, "but it's only one more game, isn't it? All this talk of yours is just rhetoric—mere appearance."

"When human beings have properly adapted to what we now are," Conrad Helier's image replied, twisting its synthetic lips into a conscientiously ironic smile, "we'll all be playing God, because it'll be the only game in town. It's not a position that can be left unfilled—not any more. We have the power and we have the time, so we have to take the responsibility."

"And where do I fit into your grand plan? Or did you only use my name in your patched-up package of disinformation to get back at me for deserting the fold?"

"You decided to be a player, Damon. We had to put you in a position where you couldn't do us any harm, for safety's sake. Anything you say about us from now on is bound to seem suspect—and I can assure you that there'll be no record of this conversation to prove that it ever took place. But if you want a place in the scheme, you only have to step aboard. Go out to L5, Damon—join Eveline and work with her. It's the place to be, nowadays and for the next forty or fifty years. After that...anything's possible, Damon. Anything's possible, for you and for anyone, if you can only cultivate the skill and find the drive."

"And that's what you expect of me, isn't it? That's what you've always expected."

"It's a natural next step. You've got the senseless violence out of your system, and you've already shed your links to that particular phase of your past, one by one. There are only two possible futures before you now: either you become a corporation man, building gaudy fantasies like this to amuse the meek; or you become a real outlaw, and a real inheritor of earth."

Damon looked his enigmatic visitor squarely in his deceptive eyes, and said: "You're not my father at all, are you? This is just one more phase of the game, one more layer of illusion." He was aware of the desperate edge in his voice, and of the fact that he was being perverse for the sake of it. As Mr. Yamanaka had said, there was a truth lurking at the bottom of the swamp of deceits, and in his own mind Damon was morally certain that he had reached that truth. However false this appearance might be, and however absurd its context, he was completely convinced that the voice with which this man was speaking was his father's voice, and that it was speaking as plainly and as honestly as it could.

"Does it really matter who or what I am?" the invader asked, quietly. "The specifics of the case are trivial; what really matters is the one thing that's obvious no matter who's alive and who's dead, nor who made what and why. Anything is possible, Damon, if you can only figure out what you want to do. If you want to, you can hide out in virtual space all your life; a lot of people will, now that the option's there. If you'd rather be a man than a rabbit, though, don't be ashamed of trying to play God in earnest, because there really isn't any other game worth playing."

"I don't owe you anything," Damon said, equally quietly. "Whoever you are, or pretend to be, I don't owe you anything. I'm my own man."

"You owe it to yourself to be and do everything you can," said the person wearing Conrad Helier's face. "Some day you will. All I'm asking you to do is to start now instead of leaving it until the day after tomorrow."

With that, Conrad Helier's simulacrum turned, and walked away into the imaginary wilderness—into an unreasonably vivid forest, like none that had ever been seen on the face of the Earth.

Brightly-colored insects fluttered into the virtual space he had vacated, impossibly pretty and precious.

A semi-human dryad stepped out of the bole of one of the trees, blinking in the sudden sunlight. She wore Diana Caisson's face, but there was nothing of the real Diana in her make-up; she was only a phantom, like the insects and the trees.

Like the invader before her, the dryad soon disappeared into a riot of color and confusion—but Damon knew before then that, whatever he chose to do today and tomorrow, he would never be rid of the challenge that had been set before him.

As his father had said, the only thing to be decided was when he would begin.

CPSIA information can be obtained at www.ICGtesting.com
224074LV00003B/110/P